Biodynamic Preparations
Around the World

Ueli Hurter has completed agricultural apprenticeships in his native Switzerland and in Germany and France. He has served as President of the Swiss Demeter association and as Swiss spokesman to Demeter International. Hurter has been involved with the Agriculture Section at the Goetheanum in Switzerland since 1994, and became co-head of the section, alongside Jean-Michel Florin, in 2010. In addition to his work at the Goetheanum, he is co-manager of the enterprise Ferme de L'Aubier, a Demeter-certified farm, bio-hotel and bio-restaurant. He also sits on the executive board of the International Biodynamic Association.

Biodynamic Preparations *Around the World*

Insightful Case Studies from Six Continents

Edited by Ueli Hurter

Translated by Bernard Jarman

Researched and written by Ueli Hurter, Dr Ambra Sedlmayr,
Anke van Leewen, Johanna Schönfelder, Dr Maja Kolar and Dr Reto Ingold

First published in German as *Biodynamische Präparatpraxis weltweit:
Die Fallbeispiele* and in English as *Worldwide Practice of Biodynamic Work:
The Case Studies* by Sektion für Landwirtschaft am Goetheanum, Dornach
in 2018. This edition published by Floris Books, Edinburgh in 2021
Text and photographs © 2017 Sektion für
Landwirtschaft am Goetheanum, Dornach
English version © 2017 Sektion für
Landwirtschaft am Goetheanum, Dornach

All rights reserved. No part of this publication may be
reproduced without prior permission of Floris Books, Edinburgh
www.florisbooks.co.uk

 Also available as an eBook

British Library CIP available
ISBN 978-178250-614-0
Printed in Great Britain by TJ Books Limited

 Floris Books supports sustainable forest management
by printing this book on materials made from wood that
comes from responsible sources and reclaimed material

Contents

Preface	6
Acknowledgments	7
Introduction	8
1. Andreas Würsch, Sagensitz Farm, Switzerland	16
2. Christoph Willer, Pretschen Farming Estate, Germany	44
3. The NRW Group, Germany	67
4. Pierre and Vincent Masson, Les Crêts Farm, France	84
5. Antoine Fernex, Truttenhausen Farm, France	112
6. Harald and Sonja Speer, Uppmälby Farm, Sweden	132
7. The Zeeland Group, Netherlands	147
8. Carlo Noro, Labicao Farm, Italy	163
9. Angela Hofmann, Sekem Initiative, Egypt	179
10. Andrea D'Angelo, Bairro Demetria Settlement, Brazil	199
11. João Volkmann, Capão Alto das Criúvas Farm, Brazil	221
12. Devon Strong, Four Eagles Farm, USA	245
13. Chris Hull, Hohepa Homes Community, New Zealand	262
14. Colin Ross and Wendy Tillman, Seresin Estate, New Zealand	279
15. Binita Shah, Supa Biotech (P) Ltd, India	311
Afterword	333

Preface
Wolfgang Stränz

This book is an excerpt of the study on the biodynamic preparations published by the Agriculture Section of the School of Spiritual Science in 2016. This study was published under the name 'The biodynamic preparations in context: Individual approaches to preparation work – Case studies of worldwide practice' and was conducted as agricultural and rural sociological research work. In addition to the individual case studies presented in detail, there is a description of the methodology of research design, case study selection, data acquisition strategy and analysis at the beginning of the study, and finally a comparison of the case studies and a summary discussion of results. This clarifies the scientific character of the study. It is available online at www.sektion-landwirtschaft.org.

The fifteen case studies not only present the individual working techniques of the respective preparation manufacturers, but also their individual references and personal attention and development of the biodynamic preparations. The reader feels themselves involuntarily attracted, and begins to identify themselves with the persons presented, regardless of the geographical, climatic or social conditions they work with.

This is the other value of the preparation study: in addition to the global comparison of the biodynamic preparation practice, portraits are drawn of the people who stand behind their respective work. The personal access to the people making the preparations is exciting.

For this reason, this book presents the portraits of the manufacturers of the preparations, which makes for interesting reading itself.

Furthermore, a portrait of Pierre and Vincent Masson is presented which was already planned for the original version of the study, but could not be taken up due to time, language and technical difficulties.

Acknowledgments

We would like to offer our warm and heartfelt thanks to all the preparation makers, their groups and the families who have received the researchers and shared their knowledge and experiences of the biodynamic preparations with them, as well as the readers of this publication.

Thanks too to the many individuals in the various biodynamic associations who have provided contacts and support for various aspects of this research.

This research has been closely supported and overseen by the IBDC (International Biodynamic Council) in its capacity as the steering committee. Benno Otter and Dr Uli Johannes König have offered their time and participated in pilot case studies to develop a consistent research methodology. David Steiger, Simon Brinkrolf, Dr Uli Johannes König, Beate Hurni and others have taken on tasks in various 'emergency' situations. Therese Jung from the Agriculture Section has managed the project backstage and navigated her way through an excessive work load due to the occupation of her colleagues with this project. The co-leaders of the Agriculture Section, Jean-Michel Florin and Thomas Lüthi, provided feedback to the project as milestones were achieved.

Dr Petra Derkzen agreed to peer-review and scrutinise the present report and help us reduce mistakes and increase clarity before publication. Bernard Jarman has made translations and polished the English language of this report.

This study would not have been possible without financial support from a number of individuals and organisations. A grant given by Software AG Stiftung enabled us to start the project and plan the worldwide data collection. Demeter International, the Verein zur Förderung Anthroposophischer Institutionen, CULTURA GmbH and Christopher Mann made significant contributions to support the work. Any remaining costs, not covered by the main project related donations, were covered by free donations to the General Anthroposophical Society and the Agriculture Section.

We naturally take full responsibility for any errors or misinformation that may appear in the text.

Introduction
Ueli Hurter, Dr Uli Johannes König

How this project came about

The preparations form a core part of the biodynamic approach to agriculture that was founded during the Agriculture Course. It is, therefore, necessary for the biodynamic movement to periodically take up this theme and address the issues that emerge. In doing so both 'internal' aspects (issues arising within the biodynamic community) and 'external' questions need to be considered.

The most recent focus on the preparations came from outside the movement as a result of the BSE crisis and the public health regulation that was introduced in response to it in 2001, and which effectively prohibited the preparations throughout the European Union (EU). A great deal of effort was then needed – including the setting up of a dedicated office in Brussels – in order to re-establish some form of legal basis for them within the EU.

The current need to consider the preparations has its origin within the biodynamic movement itself. During the three years leading up to this study, various issues have arisen in connection with the preparations. In its certification work, Demeter International (DI) has found itself having to decide which preparations comply with the standards and which do not. In other words, some standardisation is being asked for. A less visible but very real question arises in connection with the increasing number of producers growing specialist crops, especially wine. As new groups of people begin working with them and public awareness grows thanks to media interest in new wine trends, the spotlight shifts to the preparations. As it does so a large market for the specialist production of preparations begins to open up. A third aspect concerns the different schools of thought regarding the preparations and the tendency among their very dedicated representatives of emphasising their differences rather than what they have in common.

The International Biodynamic Council (IBDC) was asked to give a clearer definition of biodynamic preparations and the way they are made.

The question was: if limitations need to be set in terms of guidelines, what aspects should be considered? Who should decide what is right or wrong? Is the creation of standards the best way of achieving a) greater clarity in the production and application of the preparations, and b) strengthening the work with preparations? Could there be other social techniques for achieving this goal by, for example, providing information on 'best practice', such as with handbooks or mandatory training? The Section for Agriculture takes the view that defining the preparations in terms of standards is of questionable value – freedom is required to work creatively with the biodynamic impulse. The question from the point of view of the Section for Agriculture is therefore not about standards but rather: How can we best focus on the unsolved mysteries of the preparations so that instead of taking hasty decisions, a space is created where involved discussions can take place and a mood of genuine research is encouraged within the biodynamic movement?

It is the particular task of the IBDC to address the diversity, areas of tension and questions living in the biodynamic movement. It is manifestly clear that while the strength of the Demeter trademark is due to the clear standards that have been set, the investigative spirit living in the biodynamic movement also needs to be cared for and developed. This reflects the twin objectives so typical of biodynamic agriculture, of engaging in production while simultaneously searching for new ways forward. That is why the Agriculture Section was encouraged by the IBDC to take up the theme of the preparations on behalf of the whole movement, and to explore current questions in an exemplary way and provide deeper insights.

Unique qualities of the preparations

The preparations have several special qualities. One of them is that their existence depends entirely on their being made and applied. The reason for this is already laid down in the Agriculture Course. Following the grandiose cosmic-terrestrial imaginations given in the first and second lectures and the delicate substance lore offered in the third, we are introduced in the second half of the fourth and in the fifth lecture, to the preparations in terms of how to make and use them. The spiritual and scientific explanations that introduce them are hardly developed. This gives the impression that Steiner wanted to inspire each person to take responsibility for, and find their own, connection to the preparations. Statements made by Rudolf Steiner around the time of the course (for instance during an address given to young people in Breslau, June 1924) confirm this. They serve to explain why a purely intellectual understanding of the preparations is to be avoided. It is in the nature of the preparations that they can only become a reality once they have been made: the preparations can only be spoken and thought about by those who have actually made them. Or, stated more boldly: only those who make the preparations can know what they are.

This personal practice requires skill, willingness to act and a feeling for the preparations. It leads to a certain qualitative approach to the preparations. The inner and the outer practical aspects interpenetrate one another more intensively than they do in the case of other work on the farm. It is not easy to work with this on both the personal level, in the movement or with the wider public. This feeling of discomfort can, on the one hand, lead to a demand for greater factual objectivity and thereby threaten the more personal and meditative relationship to the preparations. On the other hand, the personal relationship can be so overpowering that the connection with outside reality is lost. It could perhaps be expressed as follows: the esoteric and exoteric qualities of the preparations thoroughly interpenetrate one another. A formulation from one of Peter Blaser's essays on the Agriculture Course is particularly pertinent in this regard:

> The only true way to understand the preparations is by working practically with them, which means making and applying them and being fully present when working with them. And the understanding that we try to gain through our own observations, personal experiences, botanical studies and the sadly all too brief descriptions given by Rudolf Steiner, can really only serve to stimulate our own inner insights and provide an impetus for the practical work.

Previous research and experimentation

Immediately before and after the Agriculture Course the preparations were the subject of intense experimentation. During the first decades, these trials served primarily to introduce these new and unusual practices to farms and gardens without any thought of verification or proof. Up until the 1950s and 1960s numerous comparative trials were carried out, often without any replication, to show the general effects of the preparations but also to compare different application methods.

In the 1970s active and intensive academic research into the preparations was started mostly at the request of, and carried out by, diploma and degree students. These projects were often one-off in nature and generally came to an end once the students had finished their research activity. During that time, however, the Biodynamic Research Institute in Darmstadt developed a long-term programme of research into the preparations in partnership with some universities. The important results have been published in a document entitled 'Ergebnisse aus der Präparateforschung'. No detailed description of this work is therefore needed at this point.

In collating the research results of the past decades it became apparent that the preparations seemed to resist scientific examination – the results were often unpredictable and depended on the context in which they were carried out. This means that although some really significant effects of the preparations were discovered during a particular trial they did not replicate when, under similar conditions, the trial was repeated. This phenomenon of the context-dependent visibility of preparation effects has, however, led to some conclusions regarding their scientific investigation.

There is a range of basic phenomena showing the effects of the preparations. Such basic phenomena are the improvement in soil life (which also includes the often quoted increase in soil humus content) greater root penetration of the soil, promotion of soil fauna (earthworms, springtails etc.) and, in relation to plant growth, enhanced germination, strong seedling development and improved ripening are often described. With regard to ripening the effects can be observed right through into the range of nutrient substances present. The influence of the preparations on manure and compost can also be assessed as a basic phenomenon – the composting process is accelerated and the emission of odors is reduced significantly.

In many cases, the effects can go in the opposite direction too. Yields may, for instance, be increased or reduced depending on conditions at the outset. This phenomenon can be described as 'system regulation'. With this, it becomes clear that the preparations do not only work via a causal relationship to life but have a superordinate guiding role that expresses itself in a balancing, healing process.

A further aspect is the challenge presented by the extremely small amounts of substance used when applying the preparations. This is often compared to homeopathy, although this is not strictly true since the preparations are not rhythmically diluted, as is the case with potentisation, but are used directly in a highly diluted form. A completely new research approach is needed in order to explore these subtle forces.

An assessment of the preparations using conventional scientific methods is only possible to a very limited extent if at all, while the context-dependent effects of the preparations suggest that a certain openness and diversity of approach is needed. In this study, therefore, a scientific research technique will be used that can give greater credence to these qualities than the usual natural scientific experimental method can hope to do.

Project objectives

The basic objectives for a 'preparations project' were formulated in response to current issues in the biodynamic movement on the one hand and to the particular nature of the preparations on the other:
- The practical work with the preparations and how it is currently being carried out should be recorded and objectively mapped out. This means carrying out field research throughout the world and making use of case studies.
- The assessment of how the work with preparations is done in various places should not consist solely of hard facts but include the person or people carrying out the work as the active creative element in the 'framework of preparation effectiveness'. Personal connections to the preparations need to be specially acknowledged.
- The many different ways of working with the classical preparations introduced by Rudolf Steiner need to be met with a positive attitude. It is evident from the above approach to the preparations that any judgments given in terms of 'right' or 'wrong' have very limited value. Every serious interpretation of the preparations is valid in terms of the living biography of the person in question.
- An objective assessment of preparation making practices, including those of a personal and scientific nature, should form the basis for dialogue within the movement. The capacity for such dialogue does not come about by itself but needs to be actively nurtured. This is accomplished to begin with by establishing a panel of cases.

Project implementation

For a long time, it was unclear who might be able to carry out such an investigation and how the agreed objectives should be achieved. Those chosen for the task had to be people familiar with the preparations and capable of entering into an open exchange with specialists across the world. But they should not be specialised to the extent that their objectivity is compromised. Dr Reto Ingold had the idea of forming a team of young researchers and entrusting them with the task. This suggestion was put into practice and Dr Reto Ingold took on the role of coaching the team.

The research team

The research team consisted of four academically trained researchers. Dr Maja Kolar is an eco-pedologist with a Master's and Ph.D. in Biological and Biotechnical Sciences at the University of Ljubljana (SLO). She is trained in biodynamic agriculture and consultancy, and is a co-worker at the Demeter Institute in Slovenia, responsible for the training of advisers and certification officers. Kolar lectures in organic and biodynamic agriculture at the Biotechnical Centre Naklo.

Anke van Leewen is a biologist with a teaching certificate for biology and art. She has carried out research work at the University of Duisburg (DE) and has practical experience on various biodynamic farms. During the project, she was a student at the Swiss biodynamic agriculture training.

Johanna Schönfelder is an agronomist who studied organic agricultural science at the University of Kassel (Witzenhausen, DE). She trained as a biodynamic gardener and farmer and gained work experience in Germany, Norway, and Switzerland.

Dr Ambra Sedlmayr is a biologist with a Master's and Ph.D. in Environmental Sciences at the University of Essex (GB), focusing on agricultural and rural sociology. She has carried out research projects in environmental studies and agricultural sociology in Portugal and England and is a co-worker for the Agriculture Section. Dr Ambra Sedlmayr took on responsibility for the scientific coordination of the project and was particularly concerned to ensure precision in the methodology and the full traceability of all results.

The IBDC confirmed the project as being a necessary next step for the biodynamic/Demeter movement. As a consequence of this relationship, the IBDC became the steering group for the project. Decisive steps and interim results were presented to the IBDC for its approval.

Ueli Hurter, representing the Section's leadership, initiated and directed the project.

The research approach is based on the methodology of agricultural and rural sociology. Constructivist agricultural and rural sociology is a scientifically recognised approach in the Anglo-Saxon world. It is not widely known in the German-speaking world, however. Broadly speaking it distinguishes itself from other scientific approaches in that objectification occurs by including rather than excluding the person involved. This is particularly relevant in the field of agriculture where the vision and motivation of the farmer are decisive in determining how nature is approached and the rural landscape structured. The participative research approach is closely related to that of rural sociology and was introduced into the work of the Agriculture Section in 2010 under the guidance of Claus Otto Scharmer and Nicanor Perlas. The methodology used in this project is thus in line with the direction pursued by the Section over a number of years. The precise steps leading to the research concept, its implementation and documentation were developed by the research team and are extensively described in Chapter 2 of the study (Methodology).

Presenting the project

The Agriculture Section commissioned and directed the study, and it is being published in the name of the Section and made available to all interested people and organisations. The study is particularly directed towards the International Biodynamic Association (IBDA), Demeter International (DI) and the Circle of Representatives of the Section for Agriculture, and especially their member organisations and their individual members. We hope that this study will stimulate interest and encourage an extensive exchange about, and reflection on, the preparations throughout the movement.

This publication contains fifteen case studies with comprehensive and easily compared descriptions showing how a selected group of people from across the world work with the preparations in their own geographical and social contexts. The methodological approach allows for an inner description to be made as well as an external one. The inner description reflects the character of the person involved and shows how the attitude, understanding, and way the preparations are made and used, is an inner process. The external view permits an assessment of the practical details of the preparations and how the original indications given by Rudolf Steiner have been individualised and adapted by the person concerned to the particular place.

The material presented may perhaps be disappointing to those hoping for sensational and unexpected discoveries. If, however, we are truly attentive in our reading, and manage to enter what Goethe describes as the 'sacred open space' and perceive the deep connection between the person involved in the activity and the universality of the preparations, it can be a deeply moving experience.

The usual practice of presenting case studies anonymously is not being applied here. The people and their intimate relationships to the preparations are open for all to read along with the spacial, temporal and social contexts within which they are embedded. We trust that it will serve to awaken an attitude of mutual respect and trust. And because our colleagues have been able to share their values and struggles in this way, we hope that this attitude will catch on and that a free and open exchange about the questions and issues involved will become increasingly possible. This would represent a significant shift away from the prevailing 'question and answer' culture and create an opening for further research and reflection on the preparations. A conscious decision has therefore been made to offer neither a comparative evaluation nor draw any conclusions from this study.

1. Andreas Würsch, Sagensitz Farm, Switzerland

Dr Ambra Sedlmayr, Dr Maja Kolar

Introduction

Andreas Würsch is a farmer and member of the biodynamic preparation making group in Central Switzerland. He is well known in the Swiss biodynamic movement for his specially built preparation store and conscientious work with the preparations.

Sagensitz farm, run by Andreas and his wife, Käthi, is situated in Büren in Canton Nidwalden, to the south of Lucerne. The climate is classified as warm temperate (clima-data.org) and the region is affected by the Föhn, a warm down-slope wind peculiar to the Alps.

Dr Maja Kolar and Dr Ambra Sedlmayr visited the farm on September 24 and 25, 2014. On the first day they joined in with the group's autumn preparation making. The day afterwards the in-depth interview and the interview on preparation practice were conducted with Andreas. A farm tour and a visit to the preparation store also took place on the September 25 between the two interviews.

Farm portrait

Sagensitz is a small mountain farm and though it is only about 480 m above sea level, it is steep and surrounded by the alpine foothills. Sagensitz farm covers 10 ha (25 ac) and was established some 400 years ago.

The farm has a flysch-based soil consisting of marine sediments and made up of a sequence of sandstones, conglomerates, marls, shales, and clays laid down and subsequently eroded during a period of orogeny. The bedrock is limestone. The soil is a fertile combination of sand, silt and organic matter. It is dark with a crumbly structure and very deep (there are no stones in the vegetable fields in the valley down to a depth of 2 m, 6 ft).

Postcard of Sagensitz farm (painted by L. Baltiswiler).

The highest temperatures occur in July and are on average around 17.4°C (63°F). January, with a mean temperature of –0.5°C (31°F), is the coldest month of the year. There is a significant amount of rainfall, with 1,800 mm (70 in) on average per year. Rainfall during October, the driest month, averages 70 mm (2 in) while in August, the month with the greatest precipitation, it reaches a peak, with an average of 140 mm (5½ in).

Andreas was born on Sagensitz in 1956, the eldest of nine children. The farm was managed intensively in a traditional and conventional way. Due to financial pressures Andreas' father decided to invest in a pig unit in 1963 and established a highly intensive production system. Farm intensification was at that time seen as being the best way to make small mountain farms financially viable. Many other farmers in the region took similar investment decisions. Andreas officially took over the farm in 1986 when his father sought to build a new cow barn and found that grant aid was only given to young generation farmers. Intensive pig rearing continued until Andreas decided to sell the pigs and convert to biodynamic farming. The farm has been managed using biodynamic methods since 1993.

Sagensitz is a grassland farm with goats and field vegetables. The grassy hillsides are cut for hay and in spring and autumn used as pasture for the goat herd. In summer the goats (42 head at the time of the visit) graze an alpine pasture. Various types of goat's cheese are produced in the dairy processing unit on the farm. There is a range of livestock including cattle (five cows, a bull and their offspring), two horses, two pigs, poultry and other small animals.

The farm produces a variety of vegetables and some fruit (mainly apples). Different types of salad crops, brassicas, carrots, potatoes, beetroot, peas, beans, asparagus, pumpkins and many other vegetables are grown. Vegetables and fruit (and sometimes cheese) are sold through a box scheme with deliveries made twice a week. The goat's cheese is sold mainly to local shops.

The farm is managed by Andreas and his wife, Käthi, and they employ two to three full-time farm workers.

Andreas Würsch, farmer of Sagensitz farm.

First steps in preparation practice

Andreas was the eldest son and, according to local custom, the eldest son was supposed to take over the parental farm. He was thus involved in farm work from an early age and feels he was lucky to inherit the farm.

After school Andreas trained as a farmer and obtained a Swiss diploma as 'master farmer'. During this time he learned all about the current conventional agricultural theories and practices. He was already becoming sceptical and starting to question conventional agricultural theories. It was especially when he learned about calculating fertiliser and feeding plans that he felt it did not ring true. He said: 'You know that you have to add 100 kg of nitrogen as a fertiliser so that you can take out a certain amount of protein. And if you don't do that, it won't work. That's what I've learned, right? But I noticed that this is not the case. For example, the grass grows everywhere across the Alps each year, right? And it never receives any fertiliser.'

Soon after he took over the farm, Andreas experienced a crisis relating mainly to the conventional pig rearing. He felt imprisoned by that work; he was continuously stressed about having to always make sure he had enough feed for the pigs and having to dispose of the massive amounts of slurry. He remembered feeling at the time how 'those slurry tanks were always full'. Disposing of the slurry meant spraying it on his fields and arranging a contract with another farmer to take some of it. Both these ways of disposing of the slurry were problematic. Andreas' own fields started to develop grass with very shallow roots. One day while turning the hay on the sloping hillside, his tractor kept slipping on the moist ground. It was very dangerous and was clearly connected to the over-fertilisation of his fields. As for the farmer who took some of his slurry Andreas felt that:

'My slurry goes to a farmer who would have a really good farm – were it not for my slurry. He just takes my slurry because he has been told that in theory he could do with more fertiliser. I somehow knew that these calculations were not true, and yet my slurry was going to this farmer. I felt: this is not right, something just isn't right here.'

These experiences made Andreas feel very uneasy about his way of farming. He reflected how 'the situation was growing ever more constricting for me'.

Around this time he became aware of biodynamic agriculture and of a more spiritual approach to life. Andreas' sister, who was active in promoting biodynamic agriculture in the region at the time, invited Andreas to join a reading group, attend courses, and participate in his first preparation-making day. His wife, Käthi, who had trained with an organic vegetable grower and knew about biodynamics, was another source of encouragement for him to follow up on the biodynamic approach. A further important influence that helped to broaden Andreas' perspective on life was having a child with special needs. The experiences he had with this child opened up new dimensions for him that extended beyond mere physical reality.

Andreas felt that the way he had been farming had no future, but at the time he had no alternative vision that could motivate and guide him. This new vision did not come immediately upon meeting biodynamic agriculture. It came, however, 'like a flash of lightning' when he started reading the Agriculture Course in the study group. The passages on fertilisation and how the task of the farmer is to enliven the soil 'immediately made sense to him'. Andreas said: 'From that moment on the biodynamic impulse was here. I now realise in hindsight, that it had in fact been prepared within me.'

Andreas and Käthi made rapid changes. They sold the pigs and started growing vegetables, and in 1993 they started the conversion to biodynamics. The first time Andreas had encountered the biodynamic preparations was on the preparation-making day that his sister had organised on the farm of Peter Appert. She had arranged for a group to meet there and make the preparations. This was the beginning of Andreas' work with the preparation group in Central Switzerland.

How the work developed

Andrea's attitude of 'always searching and testing' led to the steady development and refinement of his work with the preparations and an intimate involvement with them.

A very important influence and source of inspiration for Andreas was meeting Alex Podolinsky. Andreas met Alex for the first time in the late 1990s. At that time Andreas was questioning his work and wondering if the preparations were having any effect at all. He described how 'at some point Alex Podolinsky turned up and said that if the preparations are made in such and such a way then they definitely wouldn't work. I thought: In that case they are *not* working for me!' But then Alex Podolinsky gave him some helpful ideas that inspired Andreas to develop his preparation work further.

An important aspect mentioned by Alex Podolinsky was about the importance of storing the preparations in a state of moisture equal to the soil's moisture ('Erdfeucht'). This idea made a lot of sense to Andreas: 'Once again I had something that I could critically observe, and bring my preparation-making efforts to another level. I am entirely convinced now that these moist preparations are much more alive, and that everything needs to be subordinated to bringing the preparations to this state.'

Alex Podolinsky claimed that he could observe the effect of the preparations on the plants. This motivated Andreas to try and perceive the effects of the preparations himself. Pierre Masson then told Andreas about some aspects that he could look out for, such as a different tension in the leaves, an orientation of the leaves towards the sunlight, and a generally more ordered appearance of plant growth, something which is especially noticeable when compared to plants that have received mineral fertilisation.

After some five years of working with the preparations, Andreas began to find the preparation box he had been working with impractical. In conversations with Rainer Sax, who has a dedicated shed for the preparations, Andreas felt that having a dedicated place to store the preparations would be ideal for him too. A comment by Alex Podolinsky caused him to ponder 'where should the store ideally be placed in order for me to bring the preparations into this colloidal, soil-moist state? ... I want to give the preparations their own space, and take them away from the daily buzz.' This was how he decided to build the preparation store in its current place, at the edge of the farm with a little stream flowing underneath it. An anthroposophical architect designed the building, taking account of its orientation and choosing to use specially calculated angles. The preparation house ended up mirroring the shape of the steep mountain rising behind it.

Having a dedicated space to store the preparations meant that Andreas' role in the preparation group changed. Since the BSE (Bovine spongiform encephalopathy) crisis it has become more difficult to obtain the animal organs needed for preparation making, and it was not always possible for farmers to take the assembled organs home to bury them on their farm. Andreas therefore took on this work and started looking after the preparations after unearthing them on behalf of the whole group. He explained: 'The difficult moment is when the preparations come out of the soil. You then have to accompany them like little children. It's a difficult phase and I take much pleasure and care in doing it. Others are now noticing that when I do it, it works well.' Andreas added: 'it simply evolved like this because of the preparation house, and I must say I quite like doing it for this group and for the biodynamic cause as such.'

Preparation house on Sagensitz farm.

Preparation work as a path of inner schooling

Andreas described two main ways in which he works with the preparations. One of them is looking after the preparations on a day-to-day basis but without being particularly focused on the work. The other way is to take time and fully engage with them. 'This is then something completely different. I consider it more like a schooling.'

He takes time to do exercises and meditate on the preparations – especially in winter. He tries to enter into 'a condition of emptiness, yet still inclined towards the plant', an attitude of mind through which the preparation plant can reveal itself to him. When he manages to put himself in this open, receptive state he perceives an image rising in him and says, based on the example of a meditation on chamomile, 'I believe this picture is formed in me by the chamomile. But I have to put myself into a place where this can happen. And this requires practice. And I've only recently discovered this, I'm a beginner. In reality it is not that I perceive the plant, but that the plant reveals itself to me, and not in a physical way but on a spiritual level – on an etheric-spiritual level. That's how it really is. I work mainly on this etheric level, because this is the nearest, the lowest spiritual level, the level closest to the physical and the forces in it, yes.'

Andreas' understanding of the preparations happens at this level. 'Perhaps I can explain it better in relation to people. I perceive a person and I have this experience 'Ah, you are,' without saying 'You are like this or like that.' It is very similar with chamomile, 'Ah, you are chamomile.' You see and there is already a differentiation. Chamomile is not nettle.'

In considering the nature of these insights Andreas said 'What is chamomile? It is not really a feeling, it is a revelation, a perception. The perception becomes the whole, it's much more than just feeling.' This moment of insight and understanding 'disintegrates again, in daily life. I cannot conserve it. Hence it is a great joy when I'm able to retrieve it again.' Meditative work provides a great source of confidence for Andreas. Searching for words he explained: 'I have come to realise, when I do sometimes manage to work like this in winter, it gives me an incredible strength and certainty. It points towards the future. Or the new conciousness. My perception is certainly bungling but it is so strong, and points so much towards the future.'

Jokingly he said of his experiences with the preparations: 'I sometimes think that the preparations have only been created so that we humans can develop ourselves through them. I mean myself as a farmer. This is how I take it sometimes. Because the preparations do quite a lot for me when I consider what I have described.'

Engaging with the preparations has been a gradual process for Andreas. It is one in which the preparations have become increasingly important to him and are now, as he explained, finding their place right at the centre of his being. 'The whole schooling that I have described is extremely important to me. Through it I connect myself to spiritual dimensions. I notice the preparations really are a big gift to encourage me to develop myself. This is why they are my centre.' He described how the preparations came to take up such an important place in his life: 'It gradually takes over and suddenly becomes the most important thing of all ... I truly sense that these preparations form the spiritual dimension that I care for in my centre, and it is a dimension worth taking into my centre. It is a spiritual, divine-spiritual dimension that has been developed step by step. It keeps growing, it expands. I notice that these are quite simple substances, but when they start taking on such dimensions I can understand the old Demeter farmers who would say 'Yes, this is the most important thing'. I have now allowed this to mature in me so that I can also say, yes, it really is true. It really is a strong thing.'

Andreas Würsch's understanding of the preparations

The previous sections have shown how certain individuals have helped to shape the way Andreas works with the preparations. His practical experience, his connection to the soil and meditative work, have developed his understanding. But what is Andreas' current understanding of the biodynamic preparations? This question can perhaps be answered by looking at how he perceives the effects of the preparations, the qualities that are of particular importance to him, and the essential insights he has on how the preparations work. This will be attempted in the following three sections.

Perceived effects of the biodynamic preparations
General effects

Andreas has perceived the combined effects of the preparations on the soil, on the plants and on the feeding preferences of his animals. In relation to the soil, Andreas has observed that over time the soil has become darker and the structure more crumbly. He feels that it is 'more enlivened and has more strength' and that it has become more tolerant to stress, such as dryness, excessive moisture and the compaction from agricultural vehicles.

As it has already been mentioned, Pierre Masson suggested what Andreas should look for in plants in order to observe the effects of the preparations. Andreas mainly observes differences in relation to leaf tension. He notices this most clearly in the case of squashes and connects it with the application of the silica preparation. He says 'It is as if the plants are stretching themselves when they receive horn silica.' He also thinks that a hay crop which has received silica is more nutritious and is preferred by the animals. Andreas can also detect greater sweetness and a more intense taste in his vegetables, especially when their growth 'has been accompanied by horn silica.'

It is clear to Andreas that the 501 intensifies the effect of sunlight. He stressed that the 500 and the 501 are polarities within one whole and should be used together.

Effects of compost preparations and prepared compost

Andreas said he cannot distinguish the effects of individual compost preparations and, even though he has an inner image of each preparation, his understanding of their effects is drawn from reading the Agriculture Course and other relevant literature. He has however accumulated significant experiences in relation to the combined effects of the compost preparations on the composting process. He listed the following effects:

- decomposition processes are accelerated;
- if there is any smell, it is short-lived and mild;
- the warmth processes are more balanced, and not so extreme;
- there is a greater buffer between nitrogen and carbon, there is some compensating and balancing between these elements;
- the end result is a very dark, crumbly, living, moist compost, that smells of humus and has a good water holding capacity.

This prepared compost is 'a higher level pre-made soil that brings a new dimension of information to the permanent humus contained in the soil. It brings new life.' Andreas explained how over time he can observe that as a result of environmental stress the soil 'flattens' and loses its strength and vitality. When compost is added he senses that the power of the soil is being replenished. This refreshed and enlivened soil is then 'better able to serve the plant and enable it to develop in the way it is meant to.' He described the effect of prepared compost as giving 'a happy, balanced growth.'

A particularly important observation for Andreas is that 'with the preparations, fertilisation dematerialises; I need ever less fertiliser … I obtain good productivity while using around 50% less nutrients than one would normally say I would have needed to add.' At the same time the quality of the products is enhanced, because they are less watery than they would be if mineral fertilisers were used. Andreas, while remaining cautious about this estimate, concludes by saying: 'I would say that the preparations increase the efficiency of our plants by some 50%.'

Quality aspects

Andreas invests a lot of care in sourcing high quality ingredients for making the preparations. Because this work is being done together with the preparation group, however, he cannot and does not want to impose his own vision of quality on the other group members. The focus of his attention is therefore on the storing of the preparations, which is something totally under his management having built the preparation storage house.

The insights of Alex Podolinsky were pivotal in shaping Andreas' understanding of what is needed to maintain high quality preparations in the store. His aim is to keep them in the most enlivened state possible: 'I might become aware for instance that the nettle is not yet in a condition of full vitality, and so I ask myself: What do I need to change? It is like this with all the preparations. Each preparation has its own story, one tends to be too moist, another has a tendency to dry out.' According to Andreas, the condition into which the preparations are brought during the first stage of storage, and which are then maintained throughout storage, is of utmost importance. If the preparations reach and maintain their ideal state in storage it enables them to unfold their full potential when applied. He illustrated this in the case of chamomile: 'When the preparation comes out of the soil I have to care for it very consciously. I need to treat the chamomile in such a way that it can be fully effective when it is applied to the compost.'

Andreas has a continually evolving image of what 'the ideal condition of the preparation should be.' This image is connected with the spiritual dimension of each preparation that he develops during the meditation he does with them in winter, and which reflects the ideal 'physical body' of a preparation. The preparation in store should basically have a moist, colloidal, homogeneous consistency and be able to hold its shape. The preparations will then remain in this 'living' state: 'They remain in an amazing and almost unbelievable condition of stability ... and do not degenerate or become mineralised. And I feel that this is the effect the preparation will later have on the whole process of soil formation.' Out of this Andreas formulates his objective in relation to preparation quality in the following way: 'I want to pass on this flawless organism of the preparation to the soil so that it can help the soil become an organism capable of sustaining itself ... I think this has huge consequences, even for the spiritual world.'

To arrive at this 'flawless organism' Andreas uses various practical means for achieving the desired state of stability, moisture content and colloidal structure. He avoids adding water to the preparations to increase their moisture content however. He explained: 'Any water that I might add will not have participated in this whole cycle ... It [the preparation] really is an organism that has grown. It has been in the soil, on the meadow, in the manure, in the cow horn buried during the winter... and it comes into the pot. I then arrive and pour water in! For me this is like adding a foreign body.' In his experience a minimum quantity of preparation is needed if a stable condition is to be achieved: 'When I have a small quantity of preparation it dries out very quickly. The ideal minimum volume is about one litre or quarter of a gallon, it then lasts much longer. It's like being in a group; we are more resilient when we are together. It helps.'

Making sense of the biodynamic preparations

Andreas understands the preparations as a blueprint for soil development and that applying them in an enlivened, moist and colloidal state to the soil will serve it best. According to Andreas, all the subtle processes taking place within the farm organism are guided and find orientation through the compost preparations. Not in any material way, but on a spiritual level. He described how the preparations, and especially horn manure, permeate the farm organism and participate in the whole cycle of substances on the farm: from the soil on which it is sprayed, the plants that take it up, the food consumed by animals and the passing of it through their digestion, to its return as manure and the raw material for once again creating the 500. 'Indeed,' he said, 'this preparation is part of the whole cycle of the farm organism and is at the same time its source.' The 500 can be understood as this 'source' in its special role in bringing and guiding spiritual forces and processes into the farm organism.

The farmer plays a particularly important role in the farm organism. It is the farmer who guides the natural processes by using the preparations but also guides them through their consciousness. Andreas explained: 'I'm naturally part of this; I am part of it from the beginning. My human consciousness not only accompanies the spiritual in the plant but is fully part of it and guides the whole farm organism. This is how I think it is … One should not underestimate the power of human consciousness and the path it takes.' His understanding is that 'where the processes of life begin and are not yet physically manifest, that is where the preparations take effect. My thoughts also flow to that point, as do all my efforts relating to the preparations. They are woven into this new creation.'

The social setting

Work in the preparation group

As soon as Andreas started converting his farm, he joined the preparation group in the region. At that time the preparations were made within the group and then buried in the earth on each member's farm. Everyone then cared for the further steps themselves. The sharing of responsibilities and the organisation of the group have since changed. Two major drivers for this were the new BSE related regulations limiting the use of bovine organs, and the fact that Andreas had put up a building to store the preparations.

Today, the group is made up of a core group of some three to five people who organise the preparation-making day and everything needed for it. There is also an address list of people who are interested and are invited to participate in the day itself. Some come regularly and are also involved in harvesting preparation plants and bringing ingredients. Others join occasionally without taking on any specific responsibility. Newcomers to biodynamic farming are especially encouraged to attend the meeting. The larger group also meets two or three times in winter to deepen their understanding of specific subjects, hear lectures from invited speakers and share ideas.

Even though Andreas is likely to be the person in the group who has the most interest and experience in working with the preparations, he does not want to be the 'president' or take on a clear leading role. He sees his role more in terms of being a facilitator who helps individuals on their path with the preparations by providing inspiration, giving advice on practical techniques, and by helping to develop a personal relationship and understanding of the preparations. For Andreas the most important thing is for people to have positive experiences with the preparations. With this aim in mind he also withholds critical judgements, preferring instead to create constructive opportunities for showing, for example, how the quality of a preparation plant can be assessed or how an organ can be filled. Andreas explained: 'My wish is for everyone to take on responsibility and carry out this work in a conscientious way, and because I am also a farmer, I know how difficult this is. I am of course also aware who brought in the dandelion today – it was perhaps three people from the whole group.' According to him these dandelions were not of the best quality, but he accepted that they would be used and explained his position in the following way: 'For me, quite different aspects also need to be considered. They made an effort to collect them. Perhaps I give some hints; it is always good to look at the plants together. Perhaps I will make some suggestions so that they can do it better next time.'

Even if the quality of the ingredients and the work carried out is not always up to the standard he might wish for, Andreas feels that there are many advantages in making the preparations in a group and that a special quality develops through the common spiritual striving.

From a practical point of view, sharing the work of making the preparations lightens the burden of work on the individual. Andreas also considers that 'the practical work is a schooling for everyone involved and that it is important to do this together and keep on working at it.' Making preparations together is also an opportunity for meeting other people who work with biodynamics, of joining in a common effort, and *doing* something together', with the emphasis on 'doing'. Out of this a new quality develops:

'Yes, simply being together. We are like-minded and I don't know quite what happens then when we get together ... It is something different from making an effort on my own, the quality is different. And this is a very important aspect for me ... I think that the spiritual attitudes of the various people are then streaming together, and it is not important how advanced each person is in their spiritual striving, we are all working with this impulse of biodynamic agriculture and somehow one can perceive this common striving ... For me this is very special. It's *amazing*.'

Thus, even if from a practical point of view the work is not always carried out as Andreas might wish, he is tolerant and considers the most important thing is that everybody tries to improve and develop. Andreas is passionate about the group work and spoke enthusiastically about things the group did well, like getting silica well-ground.

Social context of the farm

Sagensitz is situated on the edge of a small village in the foothills of the Alps and the local population views biodynamic agriculture and preparation work with some scepticism. Käthi described how one person drops by occasionally to volunteer on the farm and in doing so takes on a social risk (that of being viewed by the village population with the same scepticism as the Würsch's).

Andreas does not wish to publicise his fine and subtle experiences with the preparations. Nor does he want to promote the preparations as a panacea for farming. He does not believe that being trendy is an appropriate way of working with the preparations. He is very cautious and even guarded about sharing his experiences. 'I do not advertise the preparations in any way or place them in the centre. It is enough if they are the centre for me.'

Preparation practice

Work with the preparations is done as part of a group, and even though Andreas seems to carry most of the responsibility, tasks are shared out. Several members of the group take on the work of sourcing ingredients. The preparation group meets in autumn, whenever possible on a Root day.

Andreas takes account of the moon constellations when planning his work with the preparations because 'forces come down to us from the stars. When we can incorporate them, it is certainly better than when we don't. I have made good experiences of pricking out seedlings on so called good days, the roots just grew, there was no comparison! And if it is so obvious with regards to root development, then surely it must be the same with other things as well.' He is not fixed about this, however. For him it is very important that the preparation work is carried out peacefully and calmly, as he believes that the way work is done affects the results. Were he to tell his workers 'you must do it today', regardless of other work, it would cause resentment and be counterproductive. He therefore always formulates it as a suggestion: 'If you can fit it in, it would be good to do it today.'

Field spray preparations
Horn manure preparation (500)

Four slaughterhouses in the region save horns for Andreas and his preparation group. The horns are used for seven or eight years. They then start to disintegrate and are replaced with new ones.

For Andreas it is very important to obtain well-structured cow-pats for making the 500. In autumn the young grass is very lush and the manure obtained often too watery. Andreas therefore keeps two cows in the barn for a while and feeds them with hay to obtain firmer cow-pats. He believes that something which is well-formed has a different quality from something that is not able to hold its shape. The manure is mixed with a shovel for an hour by two or three people in order to obtain a homogeneous, uniform manure. Andreas said that stirring the manure is important in order to mix the cow-pats from different cows with different consistencies so that 'one new whole' is created. The manure is transferred to the horn using a wooden spatula. To fill the horns well, they are gently knocked on wooden logs to help the manure slide down into the horn.

The horns are then laid into a pit that rises slightly towards the centre. This reflects the shape of the horns, which are placed side by side with their openings pointing downwards to prevent water collecting in the horns. Each horn is surrounded by earth; the horns do not touch each other while in the ground. The place where the horns are buried is well-marked.

Cow horns are stuffed using flat wooden spatulas.

Andreas digs the horns out around Easter time. He first cleans the earth from the horns. The horns are then left for up to two days to dry a little. The preparation can then be easily removed by knocking the horns upside down on a block of wood. The 500 is then broken up. Depending on how moist the preparation is, it may need to dry a little in the shade. It is then placed into a glazed clay pot. Unglazed pots are only used when a preparation is too moist and some of the moisture needs to be absorbed. No balls are formed out of the preparation. It is left with a crumbly structure so that oxygen can enter it. The preparation is moved about from time to time inside its pot. Andreas does not want to have compost worms or earthworms in the preparation and takes them all out. Although 500 can be kept for two to three years, in practice new 500 always replaces the old one.

For stirring, Andreas puts a handful of the 500 into 40 l (10 gal) of water. On average one cow horn per hectare is used, but Andreas thinks that the quantities applied could be reduced to the amount of horns a farm organism can itself produce.

The 500 is applied when the moon is descending and standing in front of an earth sign, because this relates to soil and humus formation. Andreas sprays in the afternoons or early evenings some three to four times a year on the vegetables and two times on the pastures. The preparation is applied with a copper knapsack sprayer.

Horn silica preparation (501)

One of the group members knows a gemstone prospector and receives second grade crystals from him. These are crushed using a machine he has access to. The resulting material is then sieved and the larger pieces are returned to the machine for further grinding, until a fine powder is obtained. New horns, preferably from a biodynamic cow, are used for making the 501.

The 501 is made around Easter time. The silica powder is mixed carefully with water, taking care not to add too much water. For burying and taking out this preparation, Light- and Flower days are preferred.

The preparation is taken from the soil around Michaelmas (September 29). The final substance looks pretty much the same as it was before burying. The 501 is then stored in a jar in a place exposed to the light but not to direct sunlight. Andreas uses one pinch of the 501 in 40 l (10 gal) of water. The preparation is stirred by hand.

Andreas has experimented with applying the 501 at different times, and applies it in the mornings or in the afternoons, depending on the effect he wants to achieve. According to Andreas, applying the 501 in the morning enhances a plant's vegetative development, and so is applied to encourage growth. Applying the 501 in the afternoon furthers the processes of generative and reproductive growth and brings a warmth impulse. This can be made use of when the aim is to promote sugar formation and ripening. The 501 is applied in the afternoon shortly before the harvesting of hay, carrots, potatoes and sweetcorn.

Andreas said that it is possible to apply the 501 at the wrong moment, for example when the plants are not yet properly rooted. He once burned the strawberry leaves by applying silica. When he was more involved in fruit production, Andreas used to spray the apple blossoms with the 501 on Light- and Flower days in the afternoon in order to deliberately burn the flower pistils and thin out the apples. This only burned the flowers that were open at the time.

Stirring and applying the spray preparations

Andreas uses a copper bowl in which to stir the preparations. He used to warm the water with a wood fire, but now uses gas, because too much smoke accumulated in the preparation house. The water is warmed up to 35°C–37°C (95–98°F). When the water reaches these temperatures he switches the heating off and the water cools down again gradually during stirring. He stirs the preparations by hand and this allows him time to school himself meditatively in areas that are important to him. He said:

'Of course I am now stirring in a totally different way from before … On one level it is just the same, I have always stirred by hand, but then the main thing is the human being.' He said that his stirring had changed in as much as he himself had changed and developed as a human being. Andreas sees the hour of stirring divided up into four distinct phases:

First phase: entering into it. Andreas speaks a verse to concentrate his mind and exclude irrelevant thoughts. This phase lasts for about 15 mins.

Second phase: focusing on the intention. In this phase of stirring Andreas connects himself with the reason for applying the preparation. With the 500 he desires that spiritual forces connect themselves with the preparation and give impulses to the soil so that it can develop in a way that most closely reflects its spiritual nature. When stirring silica he focuses on the plants. While doing so he asks for clarity and guidance concerning the way he, as a farmer, needs to act if the higher objectives are to be fulfilled. Despite being of the opinion that the preparation should be stirred without pause, he makes an exception during this phase. 'Perhaps I shouldn't say this, but I do not stir continuously. I stir and then I stop, get up and perhaps make a certain gesture. The image I have while doing the gesture is that the divine spiritual comes to me or into the water. I take a step back perhaps and then start stirring in the other direction.' His experience is that this gesture strengthens his connection with the spiritual world, and hence he feels it is acceptable for him to briefly interrupt the stirring.

Third phase: focusing on the recipient. During this phase Andreas stirs without interruption and places himself in the position of the soil or the plants that are to receive the preparation. A special mood develops as he imagines the soil waiting for the 500 or the plants for the 501. He once had an experience while spraying slurry that gave Andreas a sense of what plants are craving. He had been spraying slurry on the fields one day when the pipe burst and the slurry spattered all over him. In the process some drops of slurry landed on his tongue. He discovered how totally different this liquid was from what he had imagined. 'From that moment on I understood what the plant is craving. Slurry is something very different from how we understand it. It is a highly developed form of pre-constituted plant sap.' This was the main experience that enabled Andreas to empathise with plants and feel how a preparation could be experienced from their perspective.

Fourth phase: perceiving what has developed. In the last 15 mins Andreas focuses on perceiving what has been developed during stirring and observes how the preparation has 'become a whole'. Sometimes he feels 'it's bad', sometimes 'it's really great'. But good or not he still goes out and sprays it for he believes 'it's always true and always has an effect. It is a life process, whether good or bad. It simply is the way it is. I go and spray it out.'

Andreas has meanwhile developed a feeling for what an hour is. Previously when he started working with the preparations, a bird or a bee might come and disturb him and when he looked at the clock he found that the hour was over. This does not happen to him so much anymore, but he still gets so absorbed in his work that time passes quickly and when he looks at his watch it is almost time to stop stirring.

The 501 is always applied using a copper knapsack sprayer. This allows for a selective spraying of individual cultures. Andreas uses a tractor-mounted sprayer to apply the 500 and only on the steep slopes the knapsack sprayer is used.

Horn manure is applied in droplets and silica as a mist. His experience of applying the preparations is very similar to that of the third phase of stirring. Andreas feels as though the soil and plants are saying 'thank you, I've been waiting for this.' In this way spraying becomes a new experience of being a more conscious, more mindful farmer: a mediator between heaven and earth and a caretaker of the land. Sometimes Andreas experiences the spiritual aspect of spraying the preparations very strongly, at other times there are distractions.

Compost preparations

Andreas is very precise about the parts of the plants that are to be used for making compost preparations. From his point of view it is precisely the forces contained in the yarrow or chamomile flowers that are needed, and these should not be mixed together with other parts of the plant since these would dilute the effect of the flowers. To obtain only the required plant part, he is very thorough and takes great care when preparing the plant ingredients for preparation making.

Andreas is the one who always coordinates the procedures to get the animal organs, including getting the license to use them. He has a good contact with the local slaughterhouse, which is very helpful in providing the various animal ingredients. Every year a cow from his own farm or from a partner farm is taken for slaughter and the organs needed are returned.

Yarrow preparation (502)

To make the yarrow preparation, two or three stag's bladders are needed each year depending on their size. The bladders are either bought from the nursery at the Goetheanum or obtained from hunters in the region.

Yarrow does not grow on Andreas' farm and it is harvested 'anywhere!', as Andreas reports, by members of the group. They harvest the plant preferably on Flower days, and also dry it. At the autumn meeting the individual flower heads are removed from the inflorescence, so that no stalks remain. Andreas is convinced that 'it is the forces of the flowers that we are looking for and these are diluted when bits of stalk are present. It makes a big difference whether we take the root, the stalk or the flower of a plant. The flower represents the climax of plant development and this is exactly what we want.' The flowers are then stored in big glass jars until next spring and are stuffed into the bladder at St John's tide (June 24). Before filling the bladder with yarrow, both bladder and flowers are moistened with yarrow tea. They are then transferred to the bladder using a funnel.

The filled bladder is hung up and exposed to 'the summer sun'. It is taken down in the autumn and buried some 30 cm (1 ft) deep in a humus-rich and living soil. The bladder is surrounded with fine soil, not with lumps, so that there is a good contact between the bladder and the soil. The bladder is covered with some 5 cm (2 in) of soil, on top of which a layer of pine brushwood is added. This brushwood is slow in decomposing and serves to mark where the bladder has been buried. Sometimes when it is dug out the bladder is intact, sometimes very decomposed. The preparation however can be clearly distinguished from the surrounding soil but needs taking out with great care. Andreas is considering using an unglazed clay pot to bury the bladder in as there have been problems with mice. He is, however, cautious about making this change.

Chamomile preparation (503)

Chamomile is grown on Sagensitz farm. When it is flowering, members of the preparation group are invited to help with the harvest on a Flower day. New but fully opened flowers are harvested and then dried in the shade. The dry flowers are stored in cotton bags.

The intestines are obtained from a cow either from Sagensitz or another biodynamic farm from the region.

The intestines and chamomile flowers are assembled in autumn. The chamomile flowers are moistened with chamomile tea until they are moist but not wet. Care is needed to ensure that no water drips out of the moistened flowers when squeezed.

Plucking off the flower heads of yarrow inflorescences for chamomile preparation (503).

A special funnel has been made to help stuff the intestines. Because the funnel is not quite wide enough, however the person holding the intestine needs to carefully move it so that it can be completely and tightly filled. This requires attention and care and is part of what Andreas sees as a practical training for each member of the preparation group. The aim is to achieve a good contact between the intestine and the chamomile because 'the intestine wants to embrace the chamomile,' as Andreas explains. Also, any intestines that are not well-filled end up being difficult to find again in the spring because there is too little material to show up against the surrounding soil.

The chamomile sausages are placed in the soil singly and without touching one another. The aim is to completely surround them with earth. Some pine brushwood is again used to mark the spot so that they can be found easily in spring. The preparation is dug out taking care not to mix any soil with the preparation.

Funnel used to stuff chamomile, fixed on the table using a clamp.

After taking it out, Andreas breaks up the preparation to allow air to penetrate, and a fermentation process soon starts. The preparation changes from an initial yellowy colour to brown, while the individual flower heads disintegrate completely. The preparation becomes a homogeneous lump that can be shaped. It then needs to be moved and aerated every day. Chamomile is the most difficult preparation to store: it is either too moist and attracts flies, or else it quickly dries out. Andreas controls the amount of moisture by storing it in glazed or unglazed clay pots depending on how moist it is and by regularly aerating and moving it around.

Nettle preparation (504)

Each person in the group theoretically makes their own nettle preparation. In practice, however, most people now use the preparation Andreas produces.

The nettle is harvested on a Flower day just before it starts to flower. This is usually at the end of May or the beginning of June. The leaves are stripped off the standing nettle plants from below upwards and left in the shade to wilt. They are then stuffed into unglazed clay pipes 15 cm (6 ft) wide and 30 cm (1 ft) long. Andreas had previously used the whole nettle plant and this meant there were a lot of stalks in the preparation. In the course of time the material would separate out into fine material and stalks. Since using only the leaves, the preparation has become more homogeneous, something that feels important to Andreas. He also likes to use the parts of the plant where the nitrogen, iron and silica processes are most intense. In his understanding these processes are strongest in the vegetative parts of the plant; in the stems and flowers other processes have taken over.

The filled pipes are placed one after another in a pit, forming a 1 m (3 ft) long row, and then covered with soil. The place is marked and the preparation is taken out after one year, when the nettles outside are starting to flower again. What remains of the nettle is now very small, it is like a half moon inside the pipe, but it can be taken out very easily without being mixed up with soil. The reason why drainage pipes are used is to prevent the preparation getting mixed up with soil. The preparation is usually taken out of the ground some time before making next year's batch. Andreas normally chooses a new spot each year for burying the nettles.

The nettle preparation is stored in a glazed clay pot. It is the easiest of the preparations to store and needs almost no attention.

Oak bark preparation (505)

One person from the group brings grated oak bark to the preparation-making meeting. In order to be able to produce a homogeneous preparation that can be shaped, a capacity of great importance to Andreas, a very finely ground bark is preferred. The preparation nevertheless still tends to fall apart and not become colloidal.

This is of course typical of the oak bark and Andreas 'is not totally insistent' about getting a colloidal substance.

Since BSE regulations prohibit the use of cow skulls, horse heads are used instead. Andreas said: 'everything rebels inside me against using these horse skulls, because I have in the meantime learnt that these were race horses that have been pampered and stuffed with chemicals – and this forms the basis for the oak bark preparation.' In the past he has also used goat heads from his own herd, but has found them to be too small for the effort involved.

The brain is removed from the head using water pressure. The resulting cavity, surrounded by the meninges (brain skin), is filled with dry oak bark. Andreas thinks that it is important to retain the meninges because it is 'a sign of life forces still present in the skull.' However, when questioned further about it he realised that he had adopted this practice without thinking it through thoroughly himself.

The filled skull is buried in a swampy area where there is some flowing water. There the oak bark soaks up water until it is well saturated.

The skull is taken out of the ground around Easter and cut open with a saw. This means that each skull can only be used once. The oak bark preparation is then placed in a glazed clay pot. It stores easily, although it has a tendency to dry out.

Dandelion preparation (506)

It was agreed within the group that each farm should harvest dandelion flowers, preferably on a Flower day. Andreas also harvests quite a lot of flowers. Because so many people participate, it is hard to achieve the high quality Andreas aims for. Andreas agrees with Podolinsky's idea that the dandelion flowers to be used for the preparation should not have been visited by a bee. He maintains that once pollination has taken place a new process begins and there is a decline in life forces. Andreas has observed that a colloidal preparation is more easily attained if young flowers are used than when the flowers are older.

Drying and storing dandelion flowers is somewhat difficult since they stay moist for a long time and easily get musty. Once they are dry they need to be stored in such a way that they are kept safe from moths: either in a cotton bag placed in a cardboard box or in jars.

The part of the mesentery that encloses the small intestines is used to make the dandelion preparation. If the cow from which the entrails have been taken is quite old, the mesentery may have a tendency to rip easily. Andreas believes that a healthy, balanced mesentery has no fat on it. Andreas cuts the mesentery in such a way that 'nice big balls can be made.' These are made by placing the dandelion flowers, moistened with dandelion tea, inside a piece of mesentery and tying it up tightly into a ball. When taking them out of the ground less care is needed than with chamomile. The preparation is also broken up into crumbs and moved each day to begin with. Later on it forms a compact, colloidal substance.

Valerian preparation (507)

One person from the group harvests valerian flowers around St John's day, ideally on a Flower day. Someone made the comment that 'when the flowers are ripe they fall off.' The person harvesting the flowers needs to go through the field and shake the flower heads two to three times a day. Only the corolla is used for making an extract. Andreas explained that 'only the petals are taken, not the whole flower, one wants to use this power of the petals before they fall off the plant … not the fruiting stalks.' This is another aspect Andreas has taken over from Alex Podolinsky. When asked about it, Andreas explained that he perhaps should make different valerian preparations in order to find out if this really is the best way of making it.

To produce the preparation, a jar is filled one-third with valerian petals and two-thirds with water. This jar is placed in the shade for three days. Then the essence is decanted into another jar without transferring the flowers or the deposit at the bottom of the jar into the new jar. It is important to protect this liquid from oxidation. Andreas uses a little rubber pump to suck the air out of the valerian bottles. In this way the valerian preparation can last for three to four years. No alcohol should be added to the preparation in order to preserve it, because alcohol has a destructive effect on etheric forces. The valerian preparation is stored together with the other preparations in the peat-filled wooden box in the preparation store.

When the farm had more fruit production business, Andreas would sometimes use the valerian preparation on the orchard trees to protect them when a late frost was expected.

Applying the compost preparations

The compost preparations are applied on the manure piles, liquid manure, and the compost heaps. The deep litter is also inoculated with compost preparations and so even before it is built into a heap it will have received the preparations between one and three times. The compost preparations are inserted each time the heap is turned, every two months. This means each compost heap receives the preparations about three times.

The compost preparations are inserted into small balls of compost. A thimble full of preparation is used for each ball. Andreas has a wooden basket with six divisions for carrying the balls with preparations in them to the compost heap. He walks along the heap and every 2 m (6 ft) drops each preparation ball in turn into a hole. He then sprays about 5 ml (1 tsp) of Valerian preparation that has been stirred in 5 l (1⅓ gal) of cold water for 10–15 mins over the top using a knapsack sprayer.

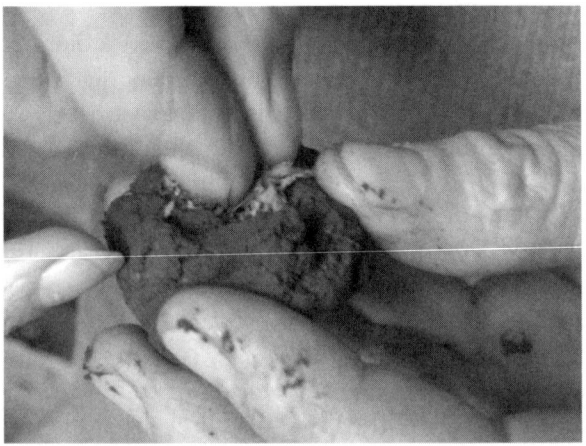

Demonstration of inoculation of compost balls.

Burying and storing practice

Burying practice

All the prepared organs, except for some of the cow horns taken home by members of the group, are buried about 30 cm (1 ft) deep on Sagensitz farm. It is important that the stuffed organs are surrounded by soil and are not touching one another. The stuffed organs are buried directly in the soil. Pine brushwood placed over a 5 cm (2 in) covering of soil, makes finding the preparations again in spring easier.

Storing the preparations

Storage is a very important aspect of the way Andreas works with the preparations. He has a purpose-built house for the preparations. Inside it is a large wooden box for storing the preparations. It has double walls containing a layer of peat. The box is sunk into the soil, so that its lid is at floor level. There is a hand operated winch attached to the wall that is used to open the heavy lid of the preparation box.

Each preparation is stored inside a glazed pot and occupies one compartment of the preparation box. In their first year the preparations need quite a lot of care. From then on they are quite stable and can last for a long time. In practice, however, they are replaced every year as Andreas feels that it is important to keep making the preparations and learning through the process.

The preparation storage box: sunk into the floor of the preparation house.

Derived preparations and other applications
500 prepared (500P)
Andreas makes the prepared horn manure preparation (500 P) because other members of the preparation group like to use it. This is made from using the traditional horn manure preparation (500) and the six compost preparations.

Cow Pat Pit preparation (CPP)
The CPP is used in the cow shed, on the deep litter, and is added to the slurry almost every week. Before application, the CPP preparation is stirred for about 20 mins in water. Horsetail tea is not in use at the moment, but it used to be applied in the fruit orchard.

Summary

There are three essential features in Andreas' path and approach to the preparations: his connection to the question of soil fertility, his social awareness, and his spiritual research.

His connection to the soil and concern for soil fertility runs like a scarlet thread through his development. Already during his training he became suspicious of the conventional agricultural theories relating to soil fertility; his experience with having too much slurry to dispose of and realising its effects on the soil made him seek an alternative way to farm, and what really moved him were passages in the Agriculture Course about the task of the farmer to enliven the soil. He now sees the preparations as being a blueprint for soil formation.

His social awareness is evident in the numerous reflections and careful considerations given to finding the best way of communicating his approach to members of the preparation group. His aim is to encourage every person to have positive experiences with the preparations and develop their own work. To this end he is even prepared, at least temporarily, to sacrifice some of the quality aspects he holds dear. It is important to Andreas that each person takes on responsibility for work with the preparations. He does not want to be the 'president' of the preparation group, but would rather work in such a way that every member can take responsibility for the work together.

The third special feature is that through his work with the preparations Andreas has been able to follow and develop his own path of inner schooling. This path relates on the one hand to the practical work and the social skills needed for group work. On the other hand side it consists of an inner journey based on his meditative work with the preparations. Andreas frequently said that he is always 'questioning and trying out', and that this is what has brought him forward in life. This attitude, together with some interesting questions and impulses given to him by other preparation makers, set him off on a path of discovery and inner development. As a result of his meditative work Andreas is modestly finding a way of developing an inner picture of the beings of the preparations. It is these inner pictures that guide the way how he looks after a preparation in storage.

According to his understanding, the preparations work on an etheric level, at a level of existence before beings and forces become manifest in matter. It is the same level as the level on which human thoughts exist. This is why our thoughts can have an effect on the preparations. The human being and the attitude with which they work with the preparations can have a profound effect on the results, Andreas believes. No matter what has been given to the preparations in this way during their process of becoming, Andreas works with them as they are, for better or for worse. For him, what has become the preparation is a truth in itself. A non-judgemental, accepting and open attitude is part of the way Andreas works.

As regards preparation practice, we can highlight the strong influences of Alex Podolinsky and the great importance Andreas places on the storage of the preparations. Because Andreas carries sole responsibility for storing the preparations, it is here that he can perhaps realise his vision for achieving the highest possible quality. For him the preparations are like organs, and as in an organism they are able to self-regulate and remain alive and healthy once they have been brought into the right condition, instead of being subject to the usual processes of decomposition or mineralisation. He believes that the ability to hold a certain form reflects a higher level of organisation.

The work in the preparation group appears to bring together practical and personal development, and a common spiritual striving for the greater good, something that feels palpable to Andreas and of huge value for the individuals concerned and for the spiritual world.

2. Christoph Willer, Pretschen Farming Estate, Germany

Anke van Leewen, Dr Ambra Sedlmayr

Introduction

Christoph Willer is responsible for the preparation work on the Pretschen farming estate in Brandenburg, Germany. He has been working professionally with the biodynamic preparations for the last 20 years. During this period he has been responsible for preparation work on two different farms. Christoph is a leading member of the International Biodynamic Preparations Working Group. He is a co-founding tutor of the East German biodynamic training course and also teaches on the north German biodynamic training course.

The Pretschen estate in the village of Pretschen lies to the south of Berlin in the Spreewald nature reserve. It has a continental climate and can be very cold in winter. Average annual rainfall is around 550 mm (22 in). Average temperatures range from −0.8°C (30°F) in January to 18.9°C (66°F) in July.

The farm visit and collection of data was carried out by Anke van Leewen und Dr Ambra Sedlmayr on April 9 and 10, 2015. After a visit to the main farm buildings led by the farm manager, Sascha Philipp, an in-depth interview with Christoph Willer and a visit to his preparation store took place. The following day Christoph was interviewed about his preparation practices and then the researchers joined him in digging out the 500. In order to gain a better overview of the big farm, various places with relevance for Christoph's preparation work were visited, such as a dandelion meadow and the banks of a brook where valerian is gathered.

Farm portrait

The Pretschen estate covers 800 ha (1,975 ac) and lies roughly 40 m (131 ft) above sea level. The soil is extremely sandy (soil value index of 24) and the large fields are surrounded by pine forests. At the same time the water table is close to the surface and there are numerous areas of low-lying bogland. It expresses an agronomic contradiction in having a water-rich landscape with many streams, and at the same time sandy fields with soil that rapidly dries out.

Barley field on the Pretschen estate.

The Pretschen estate was once a manorial estate. During the period of the German Democratic Republic (GDR), some sections of the farm formed part of a collective farm (LPG) while other areas were designated as so-called 'people's holdings' (Volkseigenes Gut, VEG) run directly by the state. After the iron curtain came down, ownership was transferred to a trust but very few resources were invested in the estate. The farm was managed conventionally until the present farm manager, Sascha Philipp, was able to take over the property in 1999 with the help of the Edith Maryon Foundation. Having grown up on a biodynamic farm, Sascha Philipp immediately set about converting it to biodynamic agriculture. He sees himself as an independent entrepreneur and is a political spokesperson for organic agriculture in the region of Berlin-Brandenburg.

The farm has around forty employees and runs four distinct enterprises. These include dairy production, cattle farming, chicory production, arable crops and intensive horticulture in glasshouses.

There are 270 milking cows and their followers making a total of 650 cattle on the farm. In 1999 the cow sheds were converted to loose housing. The milk is sold for processing to the dairy of the eco-village of Brodowin.

Only a small amount of the chicory is home grown, most of the roots are bought in from Holland and the farm focuses on the forcing operation. Rye, barley, lupins, lentils and linseed are grown in the arable section as well as fodder crops. The soil is too poor for growing wheat or spelt. An effort is made to spread manure or compost on as many of the fields as possible.

Vegetable growing, primarily cucumbers and tomatoes, is the youngest branch of the farm and was established in 2011 in response to a request from the Berlin market. The entire area of 2 ha is under glass. In the 1970s the landscape was cleared and denuded of hedges and so the ecology of the farm is currently being re-balanced with the creation of hedges and wetlands.

First steps in preparation practice

Christoph Willer was born in 1958 in north Hesse, Germany, and grew up in a rural environment near the city of Kassel where he made an early connection to agriculture. At that time virtually everyone in the village was involved in farming. While he himself lived in a vicar's family, he frequently helped a neighbouring farmer family who had children of a similar age. He described this as being the best experience of his childhood.

After completing his Abitur (roughly equivalent to a Baccalauréat), Christoph studied landscape design. His passion for ecology led him to study the natural sciences, but they left him dissatisfied and he had the feeling that something else was needed. When he was 24 years old, Christoph first came across biodynamic agriculture. In 1982 he visited a friend in Sweden who was taking a one-year course in biodynamics, and had his first encounter with the preparations and the biodynamic approach. He was immediately enthused and took countless notes. As he neared the end of his time at university he pondered how he might best direct his professional life towards biodynamic agriculture. When the opportunity of a landscape design project on a biodynamic farm arose in the context of his studies, he immediately took up the challenge and also decided to read the Agriculture Course for the first time. Christoph remembered how he sought a direct practical means of approaching the preparations: 'I read it and thought: I don't understand this, I must try it out.' In 1983 he made his first attempt at making the 500, and in 1985 joined the Study Year of the Natural Science Section at the Goetheanum with Jochen Bockemühl. This opened up the world of anthroposophy to him. Apart from Jochen Bockemühl he was also taught by Georg Maier. They helped Christoph to develop the conceptual background that was to be the foundation of his work with the preparations.

Returning to Berlin, Christoph decided to join the West Berlin Biodynamic Group and together they managed 10 ha (25 ac) of woodland. 'Dying forests' was the overriding concern of the ecological movement during the 1980s and a key objective of the Berlin group was to assist the trees using biodynamic measures.

Christoph also made the preparations with the group and was able to learn from Elmar Fournes, who was the person primarily responsible for them. From him he learnt the practical skills involved and adopted his open-minded dedication to the preparations. Christoph had to come to terms with the challenge of using animal organs. It was a thorough study of animal anatomy that ultimately opened the door to this field for him. He described how it awoke in him wonder and respect for the processes of life: 'I suddenly saw the wisdom, the amazing creative power of nature without any human input. I am sure that without this study I would never have developed such a respect for cows.'

His work in the woodland schooled his faculties of observation. Regular visits to the woods took place in order to observe changes and determine further treatments with the preparations. Christoph was by then convinced of the effectiveness of the preparations: 'We became aware, having sprayed the preparations in the woods, that our feet no longer slipped so easily on the ground. It was a steep 45° slope, which meant we easily slipped while carrying the heavy knapsack sprayers. At a certain moment we found our footsteps sinking in the ground and springing back so that we didn't slip any more. That was a very deep experience for me.'

While participating in the Study Year of the Natural Science Section in Dornach, Christoph got to know Manfred Stauffer, who was then the head gardener of the Goetheanum estate, and came into very stimulating exchanges with him. Between 1993 and 1995 they visited one another in order to learn from each other's work. Manfred Stauffer was at the time particularly concerned with the health of plants being grown for seed and sought to discover how the interaction of manure and plant compost might strengthen those plants grown for food and seed respectively. Christoph found the key to understanding this research question by reading the Agriculture Course and discovering the polarity between nutritional value and reproductive capacity. He explained: 'The plant has two properties, namely that it can reproduce itself and that it can serve as a source of nourishment for the higher kingdoms of nature. It has the tendency to develop only one of them and yet it is important to consider both aspects. I was shown how these two threads – nutritional value and reproductive capacity – can be traced right through the Agriculture Course.'

Having spent ten years working voluntarily in the woodland alongside his landscape gardening work, Christoph wanted to apply his experiences to agriculture. He was approached via the Biodynamic Group to take on responsibility for the preparations in Brodowin, the largest Demeter certified farm in Germany with 1,200 ha (2,965 ac). He worked there from 1995 until 2005, having taken over the work from Graf Finkenstein.

Since 2006, Christoph has been employed part-time by the Pretschen estate. There he has complete responsibility for the preparations. He has a great deal of freedom in the way he manages and structures the work – an unusual position to have in an otherwise tightly structured and market-orientated farm. The labour requirements for producing and applying the preparations on such a large farm are high. With some of the more labour intensive activities, like picking dandelions, he can rely on the help of colleagues. Some of the work of making the preparations is done together with the Berlin-Brandenburg Biodynamic Farm Association. If he still has time available after his preparation work, other tasks on the farm will be allotted to him.

Alongside his work with the farm's preparations, Christoph plants his own small field of potatoes where he can work more intensively with the preparations and can carry out research. He also had an orchard on Brodowin with 120 standard trees that he cultivated and where he conducted trials.

He has published many articles on the preparations and is keen to exchange experiences with the International Biodynamic Preparation Group or with Dr Uli Johannes König from the Forschungsring. Christoph is also inspired by the research carried out by Dorian Schmidt into formative forces.

How the work developed

The way Christoph has been working with the preparations on a practical level hasn't changed much over the years. The basis of his work, to which he always likes to refer back, remains the indications given in the Agriculture Course. Within the limits set by these indications, variations in preparation practice concern mainly details as far as he is concerned. He said: 'Some things do differ, such as whether the horn should lie with its opening upwards or sideways, or if the horns with silica should be closed with clay or not. There are some small things which one person will do one way and another differently.'

His teachers showed him a practical way of working which he continues with today. He believes working by hand is important, for instance grinding quartz or stirring the spray preparations.

Although he once made the CPP and produced the 500 preparation, and respects and values these developments, he feels there is no need to use any preparations other than those recommended in the Agriculture Course. He is of the opinion that 'if the indications given in the Agriculture Course are carried out well, there are very few issues regarding plant health and food quality that remain unaddressed.' He does however apply horsetail tea, stinging nettle and valerian preparations on their own and in various forms as plant conditioners to strengthen the crops. In Christoph's opinion these uses can also be drawn from indications in the Agriculture Course.

A further practical extension came about through Graf Finkenstein. He was in contact with Maria Thun and wanted to demonstrate the varying effect of the Moon in different constellations. His suggestion was to bury and dig out the 500 when the Moon was in particular zodiac constellations. Christoph continued these experiments without any intentionality of his own. He then found differences between the various 500 samples and came to the conclusion that the moon constellations do affect how the preparations turn out. Up to this day he continues making the 500 during different moon constellations. Christoph feels that there is still much to discover in this field, and he is convinced that a conscientious schooling of perception will open up new ways of investigating constellational effects.

Christoph describes his relationship to the preparations as being permeated with gratitude for the almost limitless learning opportunities they provide: 'I have discovered something which continues to be an exciting area of work for me. I am grateful too that it has a basis in anthroposophy for that is the source from which I am able to learn and from where new doors are opened.'

Preparation work is important for him since it keeps his relationship to nature fresh and alive. He said: 'I have never had such meetings with nature, or such direct impressions of her, as when I work with horn silica early in the morning; when I am outside when the sun rises, or when I spray horn manure at sunset.'

Christoph's hope for the future is for there to be many young people keen on working with the preparations. 'There needs to be people who understand and want to do it. Of course there also needs to be a context within which to exchange experiences.' This is why he is also strongly engaged with the training of biodynamic farmers in eastern Germany.

Christoph Willer's understanding of the preparations

The Agriculture Course as the foundation

Just as in his practical work, the Agriculture Course also provides the basis for Christoph's understanding of the preparations. He has read it each year for the last 25 years. Doing so has brought about in him a weaving together of theory and practice. Questions arising from daily practice or those coming from students, find illumination in the lectures of the Agriculture Course, and concepts from the course give orientation to the observations made in daily practice. They open up a way of thinking for him that he would not otherwise possess. He speaks of these concepts as 'being available to him as organs of perception.' He cannot imagine understanding a concept however without the practical aspect. He explains: 'forming a concept in your head has to do with getting hold of it and grasping it with your hands. For me these things are connected. I can only experience as evident something that I have tried or carried out myself. Truth arises as a feeling in the heart. What I perceive as being true is no intellectual decision, it is an intuitive feeling for which I must form a concept in my head.'

Christoph bases his understanding of how we gain knowledge on Rudolf Steiner's *Goethe's Theory of Knowledge: An Outline of the Epistemology of his Worldview*. He said: 'I can conceive of the whole only by holding the entire surroundings in my thoughts or perceptions. I can only perceive things when I form a concept of them. This is from *Goethe's Theory of Knowledge*: How is reality put together – through concepts and sense perceptions.' According to Christoph human beings need to become sufficiently sensitive to be of 'service to the things' in a fully conscious and ego-led way, rather than imposing their own opinions on the world.

For him, the terminology of the Agriculture Course provides an incentive for addressing the deeper questions of human existence. In this way the preparations become, for him, a path of inner schooling. He explains: 'Through the preparations we have an opportunity to become truly human.'

The question of the farm individuality

The farm individuality is an important theme for Christoph. He quotes from the Agriculture Course: 'A farm comes closest to its own essence when it can be conceived of as a kind of independent individuality', and explained that human beings need to form a concept of individuality if they are to work with a farm in such a way that an individuality can develop. Because the question of individuality lived in Christoph from the outset, an image could emerge as a concept for explaining and organising his perceptions. He is now able, after 25 years, to connect the indications given in the book *Fundamentals of Therapy*, by Rudolf Steiner and Ita Wegman, to the agricultural individuality, but also understand human beings better. He explains: 'He [Rudolf Steiner] wanted to integrate agriculture within the Medical Section and with good reason. Forming an image of the human individuality, or the individualisation of our bodily nature, helps us to structure our agricultural activity.'

The parallels between the human organism and that of the farm are for Christoph the key to understanding the preparations. He believes that the preparations facilitate the incarnation and excarnation of a number of qualities. They create a field of forces that prepare the way into or out of material existence. Christoph perceives a connection between the oak bark preparation and breathing, and between the stinging nettle preparation and the formation of blood.

He found a description by Ita Wegman in Chapter 9 of *Fundamentals of Therapy* of how the activity of silica in the organism works internally to form differentiated organs and outwardly creates the skin and sense organs. Christoph's understanding is that the activity of silica in biodynamic agriculture manifests itself not only through the 501 but also through the horsetail, dandelion and yarrow preparations. These preparations may be compared to various organ functions. He said: 'Dandelion and yarrow may be described as differentiated organs while the outwardly directed senses are reflected in the horn silica. This has given me a qualitative entry point and, I think, provides scope for further research.' The preparations make it possible for the farm to be given a meaningful structure. Christoph said: 'With the preparations we can create a human-scale landscape.'

The importance of observation in working with the preparations

For Christoph Willer the development of human faculties of perception is very important, and plant observation is essential to his work with the preparations. He observes how the plant form develops, how its leaves are placed, the colours and the moment when colour changes happen. He tries to recognise the potential in each plant and then directly support it through the preparations. With plants observed over a longer period an inner picture emerges for him. This enables him to assess at any particular moment how a crop should have developed and then support this development with the preparations. It is about bringing the plant into harmony with its surroundings and with the seasons. Using the example of how the colour of the green plant changes around St John's tide (June 24) and takes on a deeper hue, he explains: 'Up until St John's this spring green, this bursting, light-filled lime green is appropriate, but when next year's buds start forming on the trees – the same principle applies to cereals, the grains ripen to form seed for next year – at this moment the colour changes. And that needs to occur in the correct sequence.' With the help of the 500 and the 501, Christoph is able to modify the summer and winter forces and thereby lengthen or shorten their time of influence. He has had impressive experiences of spraying barley with the 501 in the afternoon to further ripening processes.

Christoph often uses terms such as devotion, humility and reverence in relation to his work with the preparations. It is clear to him that beings of the elemental world and those of the higher hierarchies are involved in making the preparations effective. Christoph tries on a practical level to train himself not only by schooling his perceptive capacities, but also by opening himself to moments of intuition. He wants things to speak to him and so places his awareness, free of all intentionality, at the disposal of the phenomena. A good example is when he climbs on to his stirring platform and acknowledges which of his preparations he is going to use. He said: 'Our time needs this devotional attitude but it does not need opinions. Who am I to judge? I have no set ideas. I am learning. I am at a stage where I really have no idea of these forces, but in working with them I allow myself to form a meaningful picture of what I need to do.'

Working with the different qualities of time

For Christoph the well-planned management of a farm includes observing the rhythmic cycles of time, which is best done in reference to cosmic rhythms. He therefore takes pains to organise his work around the Maria Thun Calendar. He avoids days that are marked out as unfavourable in the Maria Thun Calendar, both for the application and the making of the preparations.

Christoph makes a clear distinction between the astronomical constellations and the astrological signs. To his understanding, the astronomical constellations reflect the current reality while the astrological signs indicate an intention, something that is yet to happen. In working with the 500 he refers to the Moon's position in the astronomical constellations as given in the Maria Thun Calendar. But when working with the 501 on his potato research plots, it is the Sun's position in the astrological signs to which he orientates himself.

Observing the Christian festivals is for Christoph an important part of working with the preparations and of ensuring their good quality. He is consistent in only digging out the preparations after Easter, since 'in terms of fertilising the earth from a cosmic point of view, the time after Easter is different from the period before. I find that having Michaelmas as an orientation point is also important.'

Christoph has found that a farm organism only becomes receptive to the influences of the constellations once the preparations have been used over a longer period. Allowing the qualities of the different time periods to flow together is, he believes, central to the effectiveness of the preparations. He described how: 'There are different qualities of time. These varying qualities also exist among the spiritual hierarchies and the field of activity of the preparations allow them to permeate one another and cross-fertilise – to enter in and open up again but also incarnate.'

His own experience shows how relevant it is to organise the farm's 'cycle of time' (*Zeitorganismus*). Christoph explains: 'If I organise the farm in a proper way, and this also means structuring it in time, I then experience something … of an individuality. I experience something that enriches the concept of individuality. It then becomes experience.'

Human intentions have an effect

According to Christoph, even one's own intentions have an effect on the work with the preparations. Linking to the work of Bockemühl he concludes that: 'By becoming active the human being creates relationships, forms a reality that would not be there without human beings. Intention is part of the reality. The separation between subject and object does not exist for me. This of course strongly contradicts today's scientific perspective. There is no objectivity, at least not in this sphere.'

Quality aspects
The site determines the physical properties of the preparations

Christoph's preparations tend to be rather dry and the original plant structure is still partly recognisable. This is of no concern to Christoph; it arises naturally from the conditions of the site. He explained: 'The level of maturity that we see in the preparations reflects the continental climate we have here – summer drought and winter drought ... This means that the transformation processes, especially in the soil, come to a halt in summer: earthworm activity and the mycorrhiza simply stop for the summer and they have a winter break too. This means the preparations will never appear as mature as those I have seen produced in the limestone soils at the foot of the Jura hills.'

Christoph takes it as a given that the site largely determines the quality of the preparations. His understanding of the Agriculture Course means that as little as possible should be brought into the farm from outside. It follows that the preparations made on the farm are those most suited to it.

While the preparations are strongly influenced by the site on which they are made, Christoph takes great care to achieve the best possible quality.

Storage of preparations and retention of their effectiveness

Once the preparations have been taken out of the ground, their ongoing quality depends on how they are stored. Christoph's aim is for them to be in a stable condition while keeping them moist. He describes it as being an 'earth moistness'. He dries some of the preparations slightly when they come out of the ground. He believes that the preparations should not go mouldy and that handling should be carried out in a clean and careful way. Allowing compost worms to break down the 500 cannot be allowed because then the typical quality of the 500 is lost.

Because he does not wish to interrupt the living processes in the preparations, it is especially important for him to ensure that the materials used to isolate the preparations in the store are dry. If the peat surrounding the preparations is moist, the radiant properties of the preparations are drawn into the decomposition processes in the peat and are lost for further use. According to Christoph peat will only provide a protective shield if it is dry or completely saturated in water.

Christoph understands that the preparations will keep for an unlimited period. He acknowledges however that certain qualities will be lost over time and that some of the substance will dissipate.

The physical and more subtle effects of the preparations

Christoph has made various observations over the years regarding the effects of the preparations, although he emphasises that work with the preparations does not lie within what can be weighed and measured. He does not look upon the preparations as a 'repair kit' to overcome gross mismanagement. It is important to cultivate the soil using wise agricultural practices. He said: 'Sometimes mistakes are made with regard to cropping sequences or with manuring; these cannot be simply solved with the preparations. That must be accepted for what it is. Sometimes, however, it is still possible to do something with the preparations.'

The first observation he made of the effects of the preparations was a change in the structure of the woodland soil. He suddenly discovered he no longer slipped while he was working. A comment was made that on the Pretschen estate there was hardly any wind erosion, despite most of the surrounding farms – including organic ones – being affected. As well as observing plant life, Christoph watches how the animal world appears to respond to the preparations. In the case of bees he saw how they seemed to be drawn by the effects of the preparations. He described his experience with bees: 'I can say with some certainty that when I have sprayed the last field with horn silica, when the tapestry is complete, the first swarms arrive … Whether I have been successful or not is shown to me by whether a swarm of bees arrives.'

Preparation practice

Field spray preparations
Horn manure preparation (500)

Christoph makes the 500 using 600 horns each year. The horns come from the farm and from the Präparatezentrale Mäusdorf, the international preparation centre. They are re-used until Christoph notices that the resulting 500 is no longer of good quality, for example when the colour and consistency deteriorate. This mostly occurs when the horn has become soft and brittle, and makes a dull rather than hollow sound.

He can use manure from the farm, taking it from barren cows that are kept in a separate bay and fed exclusively on hay. The horns are filled and buried on the same day, around Michaelmas. Christoph observes certain constellations when making horn manure, he uses the same position of the Moon in the same constellation for burying and digging out. Each year he produces six different horn manures. He avoids the constellation of Scorpio, having had several bad experiences with it, and plans his work with the help of the Maria Thun Calendar.

The horns filled with manure are buried roughly 40 cm (16 in) deep not far from the farm entrance. He always uses the same site.

To apply it, Christoph uses the content of 5 horns in 500 l (130 ga) of water. With this he is able to spray 25–30 ha (60–75 ac). The 500 is first thoroughly dissolved in water in a small container before being shaken into the barrel. He would like to warm up the water but has no set up for doing so. Stirring is done by hand.

Christoph's ideal would be to spray the grassland with the 500 and the 501 after every cut. For practical reasons he often sprays the 500 in the evening and the morning and the 501 immediately afterwards. He avoids days with unfavourable constellations for spraying.

In one trial, Christoph made the 500 with manure from hornless cows. It became compact and mouldy after half a year. This experience showed Christoph the importance of using manure from horned cows to achieve good quality horn manure.

Horn manure – experiences and applications

After about half a year in storage, differences between the various horn manures become visible. They show different coloration, consistency and moisture levels. Christoph's preparation box contains 500 from different years and with imprints from almost all the constellations. Each 500 is different.

Christoph pointed out that the conditions under which they were made, especially when they were produced in the same year, are virtually the same. He is convinced that the constellations have an effect but he can't say exactly what it is. He regrets not having been able to make comparative planting trials. He has, however, collected more than 20 years of experience which he can present descriptively. He is guided by his intuition when deciding which of the preparations to use. Whether he chooses a lighter, looser variant, or one that is darker and heavier, may therefore depend on the situation that day; he can either support or balance out the given constellation. Depending on its particular situation he would treat a poor, sandy soil, as in homeopathy, with somewhat less transformed, lighter coloured 500, or else stimulate the opposite principle and use dark-coloured, humified horn manure.

He follows no strict rules apart from trying to find a 500 that is suited to the crops. He thus uses the 500 that has been made on Leaf days (Moon in water constellations) on the grassland and on fodder crops; the 500 made on Fruit days (Moon in warmth constellations) for cereal crops, and the 500 made on Flower days (Moon in air constellations) on those areas from which he plans to pick dandelions.

He has found 'Virgo horn manure', made when the Moon was in the constellation of Virgo, to be most suited to seed crops and for encouraging good root growth. He uses the 'Leo horn manure' in many different situations, finding it to have an all-round beneficial effect.

Horn silica preparation (501)

Since he began working in biodynamic agriculture, Christoph has made the 501 from both quartz and feldspar. He collects the raw materials directly from the fields on the farm, and he grinds it down using a large iron mortar. He sieves out the finer particles. The material therefore remains relatively coarse but also contains dusty particles. The somewhat coarse structure of the silica does not bother Christoph since he asserts that even the finest particles cannot dissolve in water. The ground material is then moistened with water so that it has a dough-like consistency which holds together but does not flow. This 'silica dough' is then filled into the horns with a teaspoon. It is buried at Easter and taken out around Michaelmas. He stores the 501 in the horns. The preparation made from feldspar has a marked pinkish colour and is therefore easy to distinguish from the one made from white quartz.

The different kinds of 501 in store.

To apply it on the farm, Christoph uses 1tsp of the 501 for a 500 l (130 ga) barrel and uses it to spray 20–25 ha (50–60 ac). He treats his research plots prophylactically with it at monthly intervals, choosing the moment when the sun moves from one astrological sign to the next – around the twenty-first or twenty-second day of each month. Christoph chose the potato as his experimental plant because it has a vigorous growth habit and is susceptible to disease. The effects of the preparations can therefore be readily observed. Silica is sprayed to enable the potato to remain connected to influences from the cosmos, something which is particularly important when producing seed potatoes. Prior to spraying, Christoph always checks to see if the plants are strong enough or whether they first need another treatment with horn manure.

Christoph uses the quartz-based variant during the 'season of light' in late spring and summer, while that made from feldspar is used more in the autumn on, for example, the over-wintering grain crops. Feldspar contains potassium, calcium, iron and manganese as well as silica. Christoph is of the opinion that with these additional minerals the preparation is able to give greater structure-forming capacities to the soil. Iron and manganese enhance the soil's ability to breathe, calcium to create structure. In this way Christoph wants to retain the soil's capacity to structure itself over winter and encourage the intake of cosmic forces.

Christoph Willer's stirring platform.

Stirring and applying the spray preparations

Before Christoph came to the Pretschen estate the preparations were being stirred by machine. For him, there was no question of doing this. He referred himself to the Agriculture Course where the indication is given that it is not helpful to use a machine for this work.

With the help of a carpenter, Christoph constructed a stirring platform from rough tree trunks. From there the preparations could be stirred some 4 m (13 ft) above the ground. By hanging the stirring broom high up, Christoph is able to stir 500 l (130 ga) by hand in one go. For spraying out, the stirred preparation is then allowed to run down into an old molasses barrel which has been converted into an air-blast-driven preparation sprayer.

When he begins stirring, Christoph makes himself conscious of his soul mood and tries to maintain a state of inner harmony while carrying out this work. To Christoph, the time spent stirring is important for connecting with the farm and reflecting on his life. Having sole responsibility for the preparations on the farm he often finds himself grappling with work related social issues when he begins to stir. These gradually resolve themselves while he is stirring and he is able afterwards to work positively with the situation. He said how much he enjoys this work and how grateful he is for being allowed to do it.

Compost preparations

Christoph makes the compost preparations either on the farm or together with Berlin-Brandenburg Biodynamic group.

Yarrow preparation (502)

Yarrow grows all over the Pretschen estate and the flowers are picked by Christoph from places that have not been mown. In the process not all the stems are removed, the corymbs are simply gathered as they are. Here, too, Christoph remains true to Steiner's exact words: 'When I make yarrow tea for medicinal use, I don't remove the flower petals individually, I take the umbels just as they are.' He dries the yarrow in his loft and stores it there in paper bags along with the other preparation plants apart from oak bark.

Christoph obtains the stag bladders from local hunters. He finds it important not to reward them with a monetary payment but to offer them a jar of honey in return, or bring some of the farm's chicory roots to a place requested by the hunters. There is a large population of red deer in the region and there are many places of refuge. The bladders are washed out, blown up, tied up with string and dried until they are needed. The yarrow preparation is not made every year, but as and when it is needed two or three bladders are filled to provide around 10 l (2 ½ ga) of preparation.

Before they are filled, the bladders are left to soak in warm water or yarrow tea for two to three hours so that they can regain their elasticity. At the same time, Christoph checks whether the bladders have been damaged. The yarrow flowers are moistened with yarrow tea before being put in the bladders. The bladders are buried in the garden behind Christoph's house so as to keep an eye on the site and protect it from wild animals. The preparation is dug out after Easter on a favourable day according to the Maria Thun Calendar. Remnants of the original plant material, such as stalks and calyxes, are still recognisable. It is quite transformed however and has a light colour.

Chamomile preparation (503)

Chamomile is the only preparation plant that is not found on the farm or in its surroundings. This is due to there being almost no lime in the soil. Chamomile is usually bought in by the Society of Biodynamic Farmers in Berlin-Brandenburg from the Präparatezentrale Mäusdorf; it sometimes comes from as far away as Egypt.

Christoph can obtain intestines from the local abattoir. They must first be tested for BSE before the material is released. This means the intestines may be several days old and must therefore be used as soon as they arrive. The chamomile preparation is buried after Michaelmas.

The site chosen is close to the river Spree, in a hollow where cold air collects and the snow lies for a long time, but where the sun still shines on it. The soil in that place consists of light clay.

When the chamomile preparation is dug out it is usually very moist. It then needs aerating and turning in an earthenware container every two days for up to two weeks until the preparation has the right moisture content – it has become 'earth moist'. The preparation in the store during our visit was already two years old and somewhat dry.

Nettle preparation (504)

Stinging nettle grows on the Pretschen estate along the edge of woodlands and near tree rows that separate fields. Christoph uses the upper part of the plant, which he either cuts with shears or with a scythe. Hard, stemmy material is discarded. Because he always uses the same pit, he first digs out the preparation made the previous year before putting in the fresh nettles.

After the plants have been cut they are left to wilt for a few hours. The nettles are separated from the surrounding soil with the help of peat and thin, untreated wooden boards.

Christoph feels that it is not only the application of the preparations that is enlivening but also the making of them, both for the people involved and the surrounding environment. This is physically visible, since where the nettles had been mown for making the preparation, an area of fresh dark green emerges, instead of tough, old, straw-like flowering stems that have died back.

Oak bark preparation (505)

The oak bark is not stored but is taken from the tree with a rasp immediately before use. The trees are growing on the estate.

Cow skulls are used and Christoph thinks it is important that they come from the farm. He only uses the skulls once because he is keen that they still have the brain skin attached. He needs two or three skulls each year, of which at least one comes from the Pretschen estate. The other skulls come from the surrounding area and are mostly from beef suckler herds.

There is a little pond close to the manor house where the skulls are buried amongst mud and rotting plant material. Here they are completely submerged and in anaerobic conditions. This is an important element for Christoph because in the autumn the water is enriched with saponins from the decaying plants. The memory of the landscape is thus bubbling in the water round the skull. Christoph explained that in its natural state, oak bark hardly absorbs any water. He could well imagine that down beneath the water in the skull, the oak bark feels a great lack of oxygen and that it then has to learn to breathe.

When the skull is lifted from the pond and the preparation is taken out, it smells of brain skin and putrefaction. The preparation is laid out on newspaper for a day to dry out a little before being put into an earthenware pot. When the preparation is ready the smell is largely neutralised.

Christoph links the oak bark preparation to the breathing process. He deduces this from processes that are outwardly visible as well as from the Agriculture Course. In the Agriculture Course a connection is made between the oak bark preparation and calcium, which regulates the ethereal. Oxygen is the carrier of the ethereal or life element and it enters the organism through the breathing process. He explains: 'The regulation of life takes place through the breathing process. How so? The oak bark preparation is able to tell us because it has already experienced it. And of course, it can communicate this to the organic processes far better than to us.'

The breathing process expresses itself outwardly for Christoph through the changing colour of the completed preparation. When it is first taken out it is usually black, whereas after contact with air it takes on a reddish colour. The red colour is connected with the iron content of the oak bark. Iron also plays an important role in the human breathing process.

Dandelion preparation (506)

Dandelion flowers are gathered on the farm. Christoph chooses a Flower day when the Moon is in a light constellation at the beginning of the flowering period. This is towards the end of April and the beginning of May. He makes sure that the 501 has been sprayed on the area beforehand.

Christoph obtains the organs of his own animals from the local abattoir. For practical reasons he uses the greater omentum but has experience with the mesentery too. He was satisfied with both and has been unable to discern any difference. He takes care when filling them not to make the packs too large, ideally no bigger than two fists. He wants to make sure they can be penetrated with the forces of the earth.

If the preparation comes out too moist he sometimes mixes it with older, dryer dandelion preparation. What is taken as the moisture content of the preparation is, according to Christoph, often fat from the mesentery or greater omentum. This affinity of the dandelion flowers to fat points towards the preparation's activity. Fat combines and brings about elasticity, forms skin and surfaces. The milk-like fluid in the lattice stimulates the liver while the liver, together with the gall bladder, plays a key role in the digestion of fat.

Valerian preparation (507)

Christoph collects most of the valerian flowers along a stream lying outside the farm. He gathers a small amount on the farm itself from plants that had been planted. He picks the flower heads on a warm and moist Flower or Fruit day, and later separates the single flowers from the heads with shears.

Depending on the quantity of flowers to hand, Christoph decides how they are processed. If there is a sufficiently large amount he presses out the juice using an old washing mangle. The flowers are wrapped in a towel for pressing.

If there are relatively few flowers he makes a water extraction. For this he uses Easter water which, following an old tradition, is collected from a spring sometime between midnight and sunrise on Easter morning. It is said that Easter water keeps for a long time without becoming stale. He uses a jar of a size that allows for the smallest possible water surface area, because otherwise the flowers float on top. Between 2–5 l (0.5–1.3 ga) are produced. He puts the jars on a window sill so the sun can shine on them during the day. After three or four weeks the liquid takes on a golden-yellow colour. It is filtered and bottled.

Christoph finds that the valerian preparation is best kept cool and in a glass bottle. To protect the preparation from light and air a dark brown bottle is chosen.

He is satisfied with both ways of producing the preparation – cold water extraction and pressing – and experiences no qualitative differences between them. The preparation has a warmth effect, but also an active remembering capacity. Christoph referred to the Agriculture Course where valerian is connected with the substance of phosphorous, 'so that the soil can find the right relationship to what are known as the phosphoric substances.'

Applying the compost preparations

On the Pretschen estate, part of the manure is spread directly on the fields and the rest is piled up in windrows. Christoph is only able to treat the part which has been set up in windrows with the preparations. He usually does this once and if the heaps are still there after half a year, a second time.

Every ten paces the compost preparations are inserted in the form of a five pointed star. The stinging nettle is orientated towards the south. The oak bark, dandelion, yarrow and chamomile then follow. Christoph developed this arrangement together with Manfred Stauffer. He takes as much of each preparation as he can hold in three finger tips. The valerian is stirred for a quarter of an hour and sprayed over the entire heap.

Burying and storage practice

Christoph constructed a 4 m (13 ft) high stirring platform on which the preparation storage box is kept. It can happen that during winter the preparations freeze. Peat is used as isolating material in the box, the preparations themselves are in earthenware pots. Christoph makes sure that the peat is dry so that it continues to act as a protective shield. No moisture is added to the preparations in the store.

Derived preparations and other applications

Horsetail tea

From the very beginning, Christoph has included horsetail tea as part of his work. He prepares it as a tea which he brews for at least 20 mins. He uses the tea either neat or diluted in the proportion of 1:7 in water. Christoph sees horsetail tea as strengthening the constitution and therefore uses it prophylactically. This means he applies it already during the winter on his potato patch. He tries to time it so as to coincide with the melting of snow, believing that this process mirrors that of horsetail. He explains it in this way: 'When the formative forces that create crystals combine with the fluid element – that is, at the transition point between solid and liquid – they can be really influenced by the horsetail.'

For Christoph there are four possible moments during the winter when horsetail can be sprayed, three times in succession (each area therefore receives three successive applications). These times are the beginning of November, shortly before Christmas, towards the end of the mid-winter period in mid February, and shortly before Easter. He thinks it is important to have sprayed horsetail at least once in the time before Christmas because, according to Rudolf Steiner, a common plant-mineral consciousness comes about during the twelve holy nights. He explained the importance of applying horsetail early on in the following way: 'It is very important to apply horsetail tea during this time because the plant and mineral worlds penetrate each other to such an extent that the tilth forming processes and the formative forces of plants are already active in the soil.'

According to Christoph, horsetail enhances the soil's water-retention capacity, which he connects amongst other things to the saponins it contains. Over and above this, he believes that horsetail tea generates an awareness that helps both the soil to absorb cosmic forces and plants to develop their organs.

Reflecting on these comprehensive functions, Christoph sees the boiled horsetail liquid as a fully valid preparation, and would support the inclusion of horsetail tea application within the Demeter standards. He also provided the impetus for the horsetail trials that have been carried out by Dr Uli Johannes König at the Forschungsring.

Plant conditioners

Christoph developed the practice of using valerian, horsetail and nettle as so-called conditioners to support the health of plants. He uses valerian preparation to strengthen moon astrality, horsetail tea to strengthen sun astrality and stinging nettle if the two seem to be out of balance. In his opinion, these conditioners serve to enhance the effect of the preparations. He uses them primarily on trees and usually fruit trees. For him it is not a new approach but arises from the comments made about plants by Steiner in the Agriculture Course. He said: 'Rudolf Steiner made a distinction between calcium and silica, the one connected to the moon, the other to the sun. It means that there is a moon astrality and a sun astrality.'

The valerian preparation is used to strengthen the moon astrality. It is stirred for a quarter of an hour or shaken for 3 mins using the potentisation developed for making medicines at the Wala company. He pours a shot glass full around the roots. It is used if a tree develops unusually large and dark leaves. He has also experimented with using it on oaks that have suffered a lack of water and developed crown damage.

If a tree has developed an unevenly balanced crown, that is if it is partly light and partly dark in colour, or fruit development is uneven, nettle can be applied in order to harmonise it. The nettle is used either as a liquid manure, as fresh leaves, or as preparation. If fresh plant leaves are used to this end, they should be shaken up in water for 3 mins.

So-called 'boat leaves', leaves whose surfaces don't open up properly towards the sun, are treated with horsetail.

Christoph sees a lot of scope in using such conditioners for strengthening plants, but it needs to be accompanied with careful observation. He is of the opinion that such conditioners will only work if the plants have already been cultivated in the biodynamic way.

Summary

Christoph Willer works with the preparations in a way that is adapted to the large 800 ha (1,975 ac) Pretschen estate. He stirs 500 l (130 ga) in one go and distributes it using a homemade air-blast sprayer. Despite the farm's large size he makes use of his craftsmanship skills, stirring enthusiastically by hand and grinding quartz in a mortar.

Through his many positive experiences, Christoph has developed a great trust in the indications given in the Agriculture Course and bases his approach on his understanding of it. There is no question in his mind about Steiner's classical preparations and he refrains from using any recombinations of them, such as the barrel preparation. He has a particular understanding of the effects of horsetail tea and would like to see it included as a recommended application in the Demeter standards. As well as quartz he also uses feldspar for making the 501. He sprays the 500 and the 501 on the grassland after every cut. He arrived at the use of various plants or preparations made from them as plant conditioners, through his understanding of the Agriculture Course. Christoph uses these conditioners not to replace but to support the preparations.

Understanding the differentiated effects of the constellations and the cosmic importance of the Christian festivals play a big role in the timing of his work. He believes that in future the effects of the constellations will become a major field of research. His practice of making the 500 during different moon constellations results in differences in colour, consistency and moisture content in the preparation.

Decades of observation have enabled him to train his perceptions to the point where he can perceive the effects of the preparations on the plants he knows well. He is continually working to extend his understanding and experience through reading and working on his own research plots.

Christoph's work with the preparations is embedded within an anthroposophical view of the world. Human beings play a large role within it through their intentions, deeds and perceptions. Christoph occupies himself intensely with this theory of knowledge and uses Goethe's epistemology as the basis for developing his understanding of the preparations. In his practical work Christoph continues the work of personalities such as Manfred Stauffer and, indirectly, Maria Thun, and thereby connects with the roots of the biodynamic movement.

3. The NRW Group, Germany
Johanna Schönfelder, Dr Ambra Sedlmayr

Introduction

As of 2015 there are 1,476 Demeter certified farms in Germany which together manage 72,588 ha (179, 370 ac) of land (boelw.de). One hundred and twenty of these farms are in the state of North Rhine-Westphalia (NRW). The making of preparations in NRW is mainly carried out in groups. There are six regional working groups in the state and each one is centred around a farm where the preparations are made. The regional groups come together there in spring and autumn to make the preparations. Although some farms in NRW make their own preparations, they still participate in the group meetings. The overall organisation of these events is carried by Ute Rönnebeck and Gabriele Heringhaus of Demeter NRW.

For the present research project, the work of one specific regional group of the NRW was chosen at random. Johanna Schönfelder and Dr Ambra Sedlmayr visited the Niederrhein Preparation Group in Kamp Lintfort on October 22, 2014 – the day of the group's autumn preparation-making session. The authors could observe, participate and ask their questions throughout the event. A further discussion also took place with Ute Rönnebeck about making preparations in a group.

The Niederrhein region of north west Germany has an essentially maritime climate. Winters are mild with little snow and the summers mild and pleasant. Average annual temperatures are 10–11°C (50–52°F). In the area around Kamp Lintfort the annual precipitation is on average 700–800 mm (27–31 in). The loamy, clay, and even loess type soils make this an intensive arable-, vegetable- and fruit-growing region.

Because the Niederrhein Preparation Group did not meet on an active farm, Ute Rönnebeck, as a group member, recommended Rolf Clostermann for the in-depth interview. Rolf Clostermann runs the Neuhollandshof in Wesel where he has been producing biodynamic fruit for more than 30 years.

Preparation work in regional groups

The Demeter organisation of NRW provides the structure and organisation of the regional preparation groups. It sets the meeting dates and ensures that all the necessary materials are available. Almost all of the animal organ material comes from the member farms and is re-distributed by the Demeter organisation to the different groups. Stag bladders are obtained via local hunters. The preparation plants are gathered on various farms. A timely reminder of the need to pick the preparation plants is sent out by the Demeter organisation to ensure sufficient plant ingredients will be available. Thanks to this organisational consistency, almost everything is now obtained from the farms. It is only the cow horns and the 501 that are bought in from the Mäusdorf Preparations Centre. In this way the Demeter organisation has been able to reduce its annual expenditure on plant ingredients and sheaths from €1,800 (£1,500, $2,200) to €200 (£175, $250).

As advisor and office director, Ute Rönnebeck tries to participate in the preparation-making sessions of all of the regional groups. Since a high proportion of the farm managers attend, it gives her an insight into the various farms, which issues are of current concern and who is not participating at the meeting. For Ute Rönnebeck, work with the preparations is clearly a job for the farm managers. She feels that it is important for the farmers to be fully engaged in making the preparations and also sees it as an important social process.

For Ute, a further advantage of making the preparations regionally is that, unlike bought in ones, they have a connection to the region. In this way they are a good alternative for those unable to make all the preparations on their own farm. Ute Rönnebeck does not see herself as an expert advising on the preparations. As a member of the group she willingly shares the experiences she has gained while working with the various preparation groups. For specific questions, problems and new impulses connected with the work, guest speakers are invited. Dr Uli Johannes König, from the Biodynamic Research Centre and Forschungsring, is a regular contributor.

Portrait of the Niederrhein Preparation Group

The Niederrhein Preparation Group meets on the farm of Trude Karrenberg in Kamp Lintfort. The farm is no longer in production nor is it Demeter certified anymore. Trude Karrenberg has, however, been a member of the group since 1985 (when she was farming the land) and is keen to continue to 'engage with the biodynamic approach' even without a farm. In this particular group it is not only the Demeter office but Trude Karrenberg as well, who provides significant organisational support for the meetings. She makes space and tools available for making and storing the preparations, and provides a meeting space, as well as food and refreshments, for the participants of the preparation-making days.

Ute Rönnebeck acts as group coordinator. She chairs the opening round of introductions, reports on the activities of the Demeter Association, shares information about current issues and upcoming events, and asks about developments on the individual farms. There were some twenty people present on the day of the visit. They were mostly farmers and gardeners, but also shop keepers and stall holders who wanted to understand more about the preparations so as to explain them better to their customers. Most members of the group were over fifty. It is not the case, however, that there is less interest in this subject among young people. It simply reflects the high average age of those running the farms.

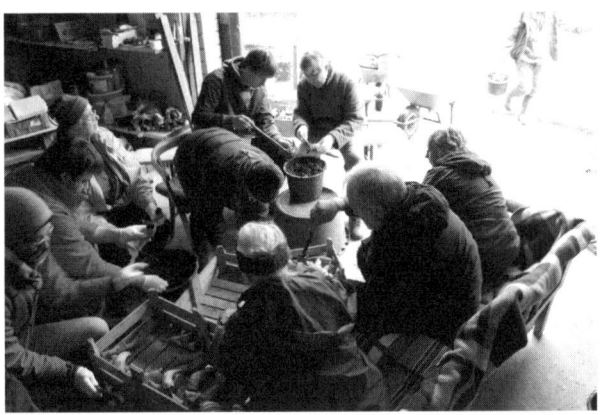

Members of the Niederrhein Preparation Group filling cow horns.

The preparation making was well prepared. On the day the interview took place, horn manure, oak bark, yarrow, dandelion and chamomile preparations were made and buried in the earth. There were plenty of materials and tools available so that members of the group could immediately set to work. There appeared to be no specialists amongst them, everyone worked as equals – no one pushed themselves forward, everyone appeared to know what needed doing. Many of the tasks – such as digging out the 501, filling horns, burying the dandelion packages – were undertaken in parallel. 'We need to it get done' was the response by many a participant when asked whether it 'is normal' to work so fast. Ute Rönnebeck also confirmed that the authors' impression of a swarm of busy bees is quite characteristic of this group: 'All the practical work must get done first if there is to be any time left for discussing and deepening our understanding of the preparations.'

How do members experience the group's work?

When asked how they felt about making preparations in a group, almost everyone had something positive to say. Again – as with the practical work – there was no spokesperson, there was simply a sharing of experiences. A recurring comment was that working with the preparations alone is hardly possible: 'No one has the time for it on their own; the ideal of a whole farm organism is scarcely achievable.' However, while 'working with the preparations regionally is still a compromise, it is a good one.' Everyone valued the group as a place for the mutual sharing of experiences, receiving advice and working together. Farmers in the process of biodynamic conversion expressed how much they appreciate the fact that as new entrants they do not have to learn and understand 'everything in one go.' They use the opportunity to ask questions and feel very supported by the group, 'one is gradually able to enter into the work.' There is also the dependability: 'one can rely on the fact that the preparations will be made and are always available – even if one is not able to participate.' The meeting itself is experienced as an 'inspirational impulse' – 'the date has been set well in advance, one looks forward to it, it gives encouragement to the work.' That there are 'no initiated ones who know it all' and everyone is able to contribute, is also appreciated. There is time to share one's own experiences with the preparations in the group and share any practical results. This helps motivate and provide inspiration.

During the in-depth interview, Rolf Clostermann did express some concern: 'Does the group encourage greater understanding or reduce individual responsibility?' On the whole, however, he felt very positive about the group. If he has questions there is always someone who can give an answer – even to an 'old hand' like himself. He sees the group as being particularly valuable for those new to biodynamic work: 'A very understandable scepticism about the whole business of preparations and the cosmos, can be quickly overcome by working alongside a group of people who are eminently reasonable and down to earth.'

Portrait of a group member: Rolf Clostermann, Neuhollandshof fruit orchard, Wesel

Rolf Clostermann runs the Neuhollandshof in the 'Lower Rhine area' – a traditional family-run fruit farm. Its soil is mostly a sandy loam, annual precipitation is 600–700 mm and the average annual temperature is 9°C.

Rolf Clostermann in his orchard.

After training as an orchardist, Rolf returned to his parents conventionally run farm in 1982 and planted 0.5 ha (1 ac) of sour cherries for biodynamic production. Pivotal in this decision were the practical results of Volkmar Lust, a pioneer of biodynamic fruit growing whom he had met at the age of when he was twenty. The biodynamic conversion of the entire farm (20.5 ha, 50 ac) took place in 1993. The farm's development was strongly focused on processing. Their own products, starting with apple juice, apple spread and apple purée, increased the range in their large farm shop. With the production of cider in 2005 processing was taken to a new level. Rolf found an answer to his question of 'how to integrate an artistic impulse into agriculture and fruit production' by planting 4,000 rose bushes on the farm. Then came the next step of flavouring the cider with roses. And so began a successful venture that has since grown larger than the core fruit-growing business itself. An alcohol free apple and rose cider produced by Clostermann can now be found in almost every organic shop in Germany. As well as regular Neuhollandshof farm tours, there are over 120 public events taking place each year in the farm's own café or in the 'culture barn'.

This enables up to 12,000 visitors per year to receive cultural inspiration and educate themselves about health and spiritual development issues. In the meantime Rolf has passed on the practical management of the orchard to an employee in order to focus on the management of the business, although he is still responsible for the preparations.

First steps in preparation practice

Rolf first became interested in anthroposophy when he was eighteen years old through a lecture given by René Maikowski. He was a Waldorf teacher in Ottersberg and had known Steiner personally. For Rolf, the intensive study of Rudolf Steiner's work is the key to understanding anthroposphy and biodynamic agriculture. His deepened grasp of biodynamic agriculture came about through his conversations with Volkmar Lust. He respected Volkmar not only for his professional knowledge of anthroposophy and biodynamics, but also his capacity for spiritual perception. Meditative practice is important to Rolf – both with regard to his own personal well being and path of inner training through anthroposophy, and to the work with the preparations. Since practising Zen meditation, visitors have frequently commented that 'something has changed' on the farm, that the farm has a 'special atmosphere.'

How the work developed
Working through observation, perception and feeling

For Rolf, working through observation, perception and feeling is fundamental to biodynamic agriculture and is something to which, as a fruit grower and in his work with the preparations, he tries consistently to adhere. His aim is to care for his fruit trees in accordance with their essential nature and following his perceptions. This means for him that they need a treatment that is different from that of annual crops.

In the early years of biodynamic management he worked a lot with prepared compost so as to reduce the amount of sprays needed. Even though he couldn't point to any visible effects, he felt that compost was the 'right medium' to work with during that period in order to stimulate vitality and enhance the immune resistance of his trees. Over the long term, however, this intensive focus on composting was no longer compatible with the ever-expanding demands of the marketing side of his business and the farm's growing emphasis on cultural activities. It was difficult to find sufficient materials in the orchard and it was too labour intensive. Today, Rolf hardly uses any compost. He gave an example of how he currently stimulates disease resistance in his trees by 'breaking off the branches instead of pruning them' and that, as a result, disease resistance and the trees' own self-healing powers have been increased.

In order to bring the influences of the compost preparations into his orchard, and following the indications given by Maria Thun, he adds some CPP to the 500 and sprays it when blossoming begins. He made this decision feeling it was right. CPP, the 500 and the 501 are regularly applied on his farm.

Meditative work and therapeutic eurythmy

Zen meditation taught Rolf to make himself 'inwardly empty'. He makes a connection here to what he understands Rudolf Steiner to be saying in *The Philosophy of Freedom*, which Rolf described by saying 'that as a therapist one needs to become inwardly free and empty so that a condition of inner receptiveness for the needs of the 'patient', the fruit tree or nature can arise and be recognised'.

He said there is a big difference whether in trying to understand a phenomenon, 'I allow the thing to explain itself or whether I try to explain it. In the latter case it is not the thing that is speaking but my inner self'.

Apart from his meditative activity, Rolf has found support for his work with the preparations through therapeutic eurythmy. The stirring and spraying of preparations has been accompanied, since 2002, by eurythmy therapist Sylvia Weyand and her partner Florin Lowndes. Rolf recounts: 'Together with Sylvia we introduced a series of eurythmy exercises to our work with the preparations on the farm. Thus, for instance, the 'halleluja' exercise developed by Rudolf Steiner to cleanse spaces – or in our case to protect against external or internal negative energies – is carried out before stirring, after stirring, and after the preparation has been sprayed. Or again the 'Evoe' exercise is done at certain places in the orchard as well as for a number of other things. Efforts have been made in connection with this work to perceive the energy points on the farm and discover how the work of elemental beings can be supported through eurythmy using verses by Rudolf Steiner.'

Rolf found a new connection to the process of stirring by working with eurythmy: 'When we spoke the Foundation Stone Meditation, the Lord's prayer or a section from St John's Gospel, I noticed that I was developing a completely new relationship to horn manure and horn silica and that I was seeing the stirring process in a quite different way than in years past ... A big moment came when I dispensed with the use of the stirring machine ... For the first time I had a very strong experience of elemental beings while stirring and then spraying out by hand. I didn't see an elemental being physically but sensed its presence behind a birch tree. It became more and more quiet and when you spray by hand you make a certain movement with the brush. At one moment I watched myself. I followed the movement and recognised it as a gesture of blessing. The sound of droplets falling on the leaves became ever louder in my perception. And then I noticed an elemental being behind a birch tree, I cannot say what it was what had been observing me. This was when I saw for the first time how strong an effect the preparations can have on the elemental world. Sudden recognition dawned on me.'

An experiential approach to farming

Nowhere according to Rolf, does biodynamic agriculture appear more strongly as a farming approach based on experience than it does in connection with the preparations. As such it needs to be practised and is, in his opinion, positively supported by spiritual work. 'I would not,' he emphasises, 'like to see biodynamic work reduced merely to the preparations.' The preparations are most of all important when there is something missing in nature. If the natural world is in balance and there is nothing missing, then according to him the preparations are unnecessary. The ideal is reached when no more preparations are needed.

At the moment he can see plenty of good reasons for using the preparations but would like to have the freedom not to do so through his own understanding. Up until now there has never been a year on Neuhollandshof when the preparations have not been applied. This is also because, during the process of stirring, Rolf finds himself being positively influenced by it. If he is asked by some random person about 'this strange work with preparations' he answers: 'Stirring preparations? I do it primarily for myself.' Everyone can after all acknowledge that meditation (as experienced through stirring), can help bring about a healthy psychic condition. 'I work with the preparations because it does me good, while at the same time having a fine and positive effect on nature. I think it is a very important aspect, and could be used to help explain perhaps why someone converting from organic to Demeter, should be using the preparations – it offers him a chance to at last be just with himself.'

Rolf Clostermann's understanding of the preparations

Working with the preparations has special significance for Rolf during the thirteen Holy Nights, beginning with the birth of Christ (December 24) via Hogmanay (December 31) to Epiphany (January 6). During this period he works particularly with the 500 and the 501. In doing so he cultivates a special awareness for the in-breathing and out-breathing processes of nature: The out-breathing process begins on January 1 and ends with the summer solstice. The phase of in-breathing then starts and continues until December 31. As the year draws to a close he has to know what should develop in the orchard during the following season. On this rests his decision as to whether a last 501 spray is applied on December 31 (in-breathing) to encourage differentiation, or if instead he applies the 500 on January 1 to encourage vegetative development (out-breathing).

501 is generally seen by Rolf as being a very important preparation, and he sprays the 501 to mark the birth of Christ. 'The preparation can, in this way, help to connect the elemental world to this event. And I think that our doing this is enormously important for the lower spiritual hierarchies to which the elementals belong.' The 501 is applied later in the season, usually at the end of June, to support the ending of the vegetative growth period. It is also applied before harvest to enhance the colour and flavour of the fruit. 500 is sprayed on newly planted trees. The choice of a specific day to spray the preparations (for example, those given in the Maria Thun Calendar) is less important to Rolf than the timing and the impulse set in motion by the preparations during the course of the year.

Stirring and spraying

Rolf has his own way of stirring and spraying the preparations: 'Stirring is carried out in the classical way by forming a vortex and releasing it. I don't calculate say 30–40 l (8–10 ga) per ha, as is often recommended. Instead I have a stirring barrel containing around 60–80 l (15–21 ga), stir it for an hour and then set off with a galvanised bucket (not plastic) and spray the horn manure using a brush, and horn silica with a knapsack sprayer – made of copper of course. That is the technical aspect. I spray everything by hand. I create something like a magic circle around the orchard, which is possible since it is contained within one area. I can easily manage it within two to three hours. I spray towards the outside as well as inside. I developed this approach in discussion and with the approval of a Demeter consultant who is one of the most well-read anthroposophists I have known – because the information contained in the preparation propagates.' He finds that what he read from Hugo Erbe also supports this approach to spraying. Rolf continued: 'Unlike with fungicide applications there is no need for each leaf to be sprayed ... And if I feel strong enough I will make another diagonal pass. I have also asked some clairvoyant people to accompany me and find out whether it is a good thing to do. And it is good. For us and for the farm individuality. Perhaps it is also connected with the spiritual work. I simply do it and no one from the Demeter inspection team has ever come along and criticised it.'

Rolf Clostermann reckons that holding to fixed stirring times is only necessary for those starting out with the preparations. He trusts his own sense for when a preparation is 'ready'; to have this freedom and ability, he asserts, requires consistent inner work. He also relies on his feeling to determine whether a verse should be spoken during the stirring.

Assessing quality and effectiveness

Rolf feels that the best way of assessing the quality of the preparations is through the personal experiences of those working with the preparations. He has difficulties with the setting of general and fixed parameters regarding the assessment of growth or taste. It is important to him that preparation quality is not approached in a factual and mechanical way but rather in terms of process. He works with and uses the preparations made by the Niederrhein Preparation Group. He considers that three factors influence their level of effectiveness: their basic quality, the way he works with the preparations (through for instance meditation, therapeutic eurythmy, devotion), and how they have been influenced by the elemental world, 'which of course will do something with the preparation that has been applied.'

The extent to which the preparations enhance the health of individual plants remains for Rolf an open question. He feels that their effects can be sensed rather than seen. He mainly experiences 'how the energy and mood of the farm changes.' Feedback from the many visitors confirms this perception when 'even a local farm woman who knows nothing about biodynamic agriculture comments on the special atmosphere of our farm.' This tells him that the entire farm benefits from the use of the preparations. A measurable change that has occurred is the increased diversity of wild flora and fauna in the orchard.

Hopes for the future

Rolf Clostermann would like the Association to support individual ways of working with the preparations instead of discussing whether to produce a fixed set of production standards. This would be best served by group sharing – especially if those working with the preparations would pay more attention to their perceptions and experiences.

Preparation practice

All eight classical preparations are made by the Niederrhein Preparation group as well as the CPP preparation. No particular account is taken of the constellations when making the preparations, the date is chosen primarily to suit the group. Individual members do however observe the Maria Thun Calendar for spraying. The preparations are made as close to Michaelmas as possible and dug out around Easter. The stinging nettle is made and dug up in June.

Field spray preparations

The horns and ground silica required for making the spray preparations are obtained from the Mäusdorf Preparations Centre. The horns are used for four or five years until they become too 'rubbery'. A horn is no longer used when its open end has become soft. The horns for the production of the 500 and the 501 are buried in the same place each year. The criteria for selecting the burial sites are practical in nature – they should not be too wet, nor too close to hedges and trees lest plant roots penetrate them. This also applies to the compost preparations.

Horn manure (500)

The manure came from two different farms and had been collected on pasture from dry cows. While the manure from one was green and somewhat slimy, the other was dry and contained remnants of the grain husks used in the feed. It was assumed that the possibly insufficient quality of the manure would be equalised in the end product by mixing the two types together. The horns were cleaned and brushed down and then filled with manure using sticks. The horns were buried with their openings facing down and each layer of horns was covered with a thin layer of soil. The reasoning behind this procedure is a 'combination of practical and inherited wisdom' according to Ute Rönnebeck. It is intended to 'protect the horns from rain water and retain the forces streaming into the horn from the earth.'

Horn manure is stirred and sprayed in the afternoon. The main crop to benefit determines the choice of day. If potatoes are the main crop then a Root day will be chosen, according to the Maria Thun Calendar, with any other crops being sprayed too.

Horn silica (501)

New horns are used for making the 501 each year. In spring the ground silica is made into a paste with water and put in the horns. The filled horns are then left to stand in a bucket of sand so that the surplus water can be poured off the surface. The horns are then buried with their openings facing downward and a board weighted with a stone is placed on top to mark the site. Eight horns are sufficient for the whole group. 501 is usually stirred and sprayed in the early morning. To help ripen the fruit and root crops an afternoon spray is carried out. The grassland is sprayed once or twice with the 501. Mains water is generally used and stirring is done by hand; only rarely is a machine used.

Compost preparations

The yarrow, dandelion and chamomile preparations are hung up under the south-facing house eaves over summer. They are protected from rain and animals by placing them in a rabbit cage. According to one group member, the decision to hang the chamomile and dandelion up in the same way as yarrow is seen as justified by the Agriculture Course where it says: 'one could do it in the same way as yarrow.' All the preparation plants destined for the animal organs are first dried and then moistened before stuffing into the animal sheaths. Only the dandelion is sometimes used fresh.

Yarrow preparation (502)

The flowers are picked and dried and then moistened with yarrow tea before being put into the bladder. The filled bladders are hung up in an airy place protected from rain.

Chamomile preparation (503)

The flowers are picked and dried and then moistened with chamomile tea, before being filled into the intestines sometime between the end of March and the beginning of May). The chamomile should only be slightly moistened to prevent it going mouldy or putrefying. The filled intestines are then hung up in a well-aerated place and protected from rain.

Nettle preparation (504)

Shortly before flowering, the whole plant is mown with a scythe, left to wilt a little, then placed directly in the soil and covered with boards to mark the site. Care is taken to make sure the site does not get overgrown with weeds to prevent penetration of the preparation by roots. If the finished preparation has a high proportion of stalky material when it is dug out, it is then broken up with either fingers or scissors. A recognisable stinging nettle structure is taken to indicate a good quality preparation.

Oak bark preparation (505)

Pieces of oak bark are ground in a nut mill and put into the cow skull. It is then placed in a large plastic barrel, covered with humus material (rotting leaves) and put under a dripping gutter. The skulls are used for three or four years and there 'may also be bull skulls' amongst them. The discussion as to whether the oak bark should be put in the skull or the brain skin is known to the group, but they consider it irrelevant when a skull is used for several years.

Dandelion preparation (506)

The dandelion flowers are picked fresh, sewn into the greater omentum, and hung up in the open air protected from rain. Dried organ material is used.

The preparation group of NRW stores the dandelion preparation dry.

Valerian preparation (507)

The group does not make valerian preparation together; several farmers extract the valerian juice on their own farms and make any surplus available to the group. The group is, however, planning to grow and process its own crop. When it is applied, a few drops of valerian preparation is added to 10 l (2½ ga) of water and then stirred by hand for 10–20 mins. It is then sprinkled over the compost. It serves to stimulate flowering and earthworm activity.

Applying the compost preparations

Slurry, deep litter manure and compost are all treated with the compost preparations. When treating a compost pile, the dry compost preparations are inserted (the amount 'according to feeling') directly into the moist compost. On small compost heaps, the nettle preparation is usually placed in the centre and the other preparations in the four corners. On longer windrows the preparations are added one after another in a long row down the middle. The holes are made with a stick. The preparations are put into balls of manure or soil. The holes are then closed again with manure. Newly built compost or manure heaps are prepared once, and slurry every two to three months. Deep litter manure is prepared by treading holes into it and adding the preparations. It is then sprayed with valerian (stirred for 20 mins) before a new layer of straw is put down.

Burying and storing practice

Burying practice

When the organs, stuffed with yarrow, chamomile and dandelion, are to be buried, a non-biodegradable string is tied to them to make them easier to find again. A layer of elder twigs is placed beneath and above the stuffed organs in the pit, to discourage mice. Then they are covered with earth. The pit is marked with posts, the same site is used every year and its location is noted on a map. When it is dug out any remaining organ material is carefully removed. The preparation is laid out to dry on sheets of paper in plastic boxes before being stored in the preparation box.

The preparation store of Trude Karrenberg.

Storing the preparations

The preparations are stored dry on many farms in NRW. This means that after they have been dug out of the ground they are dried and only then do they go into the preparation box. According to Ute Rönnebeck, dry storage was recommended by a former Demeter advisor and has since 'become established practice and is carried out as a matter of course, though many also make moist preparations.' Trude Karrenberg also dries them. The argument used by the group to justify their decision to dry the preparations is that 'storing in a moist state only works if there is someone who is able to continually monitor the preparations' – otherwise they go mouldy. They are stored in preserving jars, or externally enamelled containers, in a wooden box filled with peat. Each container is surrounded with a ring of copper. Apart from the group preparation store (to which everyone has access), most of the farms also have their own storage facilities.

Assessing quality

When the preparations are dug out their quality is discussed. According to one group member, 'there is no one in the group who can really perceive quality,' which is experienced in the group as a real lack. Criteria such as smell and colour changes during storage, as well as how well the preparations keep, are discussed. It is considered positive criterion if the original plant structure is still recognisable when it has been dug out.

Derived preparations and other applications
Cow Pat Pit preparation (CPP)

The CPP is produced according to Maria Thun's guidelines. It is especially valuable for organic farms who wish to become Demeter certified. If CPP is applied it is possible for them to gain Demeter certification after only one year in conversion.

Summary

In NRW there are six regional groups who make the preparations. The Demeter office organises and finances the work. The practical aspects of work with the preparations was the main focus of the group visited. Its organisational structure has the following advantages:

- although the preparations made are not unique to each individual farm, they do have a link to the region unlike those bought in from elsewhere;
- it ensures that the work is carried out and that a plentiful supply of preparations is available for all the farms;
- the required materials can be distributed effectively in the region and the responsibility for obtaining plant and organ materials can be shared;
- the preparation-making sessions provide inspectors and advisers with a good opportunity for meeting the farmers, gaining an up to date impression of how the preparations are being used on the various farms, and offering further support;
- working as a group serves to motivate and inspire, and above all supports newcomers and helps them find a gentle way into their work with the preparations.

Rolf Clostermann's approach to the preparations is based on his own personal experiences, observations and feelings. He therefore makes his own decision about when the process of stirring is complete and sees no necessity (on his farm) for the blanket coverage of an area when spraying. Using meditations, mantras and verses to support his work with the preparations on a spiritual level is as important to him as is feeling how the preparations are also of personal benefit to him. He believes that the preparations have a direct effect on the world of elemental beings. The direct and visible effects of the preparations on plant growth remain an open question to him. He is, however, convinced that the spraying of preparations has a positive influence on the atmosphere of the farm and supports the diversity and vitality of the plants and animals living there. He hopes that work with the preparations will remain open to individual interpretation and not become fixed by rigid guidelines.

4. Pierre and Vincent Masson, Les Crêts Farm, France

Johanna Schönfelder, Kathrin Ortlieb, Wolfgang Stränz

Introduction

In France, there are 500 Demeter-certified farms, of which 430 are in agricultural production and 70 are downstream enterprises. Seventy per cent of these farms are wine-growers. In France, biodynamic work is divided into three organizations: Demeter France, Le Mouvement de l'agriculture biodynamique (MABD) and L'Association Soin de la Terre, pour la recherche et la diffusion d'informations (Association for care of the Earth, for research and dissemination of information). Demeter France is responsible for certification; MABD is the body for information, education and training.

L'Association Soin de la Terre organizes and operates mainly biodynamic research at national and international level.

At Les Crêts, a small country seat in Burgundy, France, Pierre Masson and his son Vincent produce large quantities of preparations. Around 60,000 horns are filled annually for the production of horn manure preparation and a large amount of compost preparations are produced. In addition to the importance of the Massons as large manufacturers and distributors, they are recognied for their international consulting and research work. Their company, BioDynamie Services, managed by Vincent, distributes various dried plants for the production of herbal teas and decoctions, and carries out advanced training for those who want to develop a precise method of preparation making. In addition, Vincent works as a consultant on behalf of BioDynamie Services and gives basic and advanced courses on biodynamic work.

From October 2–3, 2014 Johanna Schönfelder and Kathrin Ortlieb visited Les Crêts. During their visit, horn manure was produced. The authors were able to observe this process directly and could thus gain an impression of the preparation work of the Massons. The interview on the preparation work was conducted during a guided tour through all the stages of the preparation production. Pierre and Vincent were the interview partners. In addition, insight into the work of the Massons was provided by means of photographs presented by Pierre and films of their consultation and publications.

On April 20, 2017 Wolfgang Stränz visited the Massons to clarify questions that had remained open due to technical and language problems. The recovery of the horn manure preparation had already begun, and he was able to gain an impression of this work

Since Les Crêts is not a farm, but there is a close cooperation with the nearby Demeter-certified Domaine Saint Laurent and other biodynamic farms as well, there was a guided tour for the authors lead by Vincent. On the biodynamic farm of 135 ha (334 ac) one can observe vegetable cultivation, arable farming and dairy cattle. There is a dairy, a bakery and direct sale. Domaine Saint Laurent is the main supplier of the cow manure needed by the Massons.

Vincent and Pierre emphasize that they have a common position with regard to the preparation work and are always in a fruitful discourse with each other. Therefore, in the following, unless the information or quotation is explicitly person-bound, they are usually quoted as 'the Massons'.

Farm portrait

Les Crêts is a collection of buildings for residential and work purposes on an area of 3.5 ha (9 ac). It is secluded and solitary. Les Crêts was purchased by Pierre and Florence Masson as a ruin and redeveloped. The complex comprises two residential buildings, a guest house and a building for BioDynamie Services. In the centre of the farm is the workshop for the preparation work. This consists of a cellar for storing the preparations, a covered workplace for the production, as well as a hall in which the preparation plants can be dried and stored. Here dried organs as well as working materials have their place.

In the meantime, the construction of the buildings has been expanded considerably. In addition to offices for the management of the farm, and storage and packing rooms for the dispatch of the preparations and the dried plants, a laboratory has been created for the production of soil chromatography and soil analyses. In addition, there will also be seminar rooms for advanced training courses of preparation making and further storage facilities for the preparations. Overall, there was a buoyant building activity and the size of the buildings will more than double.

On the small pastures of Les Crêts cattle of the Domaine Saint Laurent are grazing. The Massons themselves keep chickens as they find the animals invigorate the place. Les Crêts is situated on a hill, with views of the valley and the hills opposite. The place has its own character with various seating areas. The terrain is terraced and offers space for vegetables, herbs and flowers. These – as well as the residential and working buildings built out of the regional granite rock – fit well into the landscape. The preparation work is visible everywhere and you can enjoy the beautiful view at the workplaces. Even the stuffed stag bladder has a rain-protected panoramic view from the balcony. Everything is clean, bright and tidy, inviting, noble and fine – and yet simple.

Les Crêts is located in Burgundy near the small town of Cluny. Burgundy is known as a wine-growing region. Another focus of agriculture is livestock farming. The landscape is characterized by small-scale farmland, gentle hills, forest and numerous rivers. The climate is very diverse, influenced by continental climate. In the area of Les Crêts, there are warm summers and dry-cold winters. The average annual temperature is 11°C (52°F), the average minimum is 6.6°C (44°F) and the average maximum is 15°C (60°F) per year. The annual precipitation is 700–800 mm (28–31 in). The subsoil around Les Crêts is dominated by sandstone. The soil type is predominantly granite and the pH of the soils is 5–6.

The preparation work on Les Crêts is shared by Pierre and Vincent. Vincent manages the practical production and supervises the storage and the quality of the preparations.

In the meantime, for reasons of age, Pierre is stepping back more from practical work. Vincent is also trying to transfer his responsibility for the manufacture of the preparations to his employees. Through new corporate and social forms, the Massons hope that their goal to produce the best possible preparations will also be expressed through greater participation of their employees. As a result, the company hierarchies should become flatter.

Six permanent employees and up to ten seasonal workers mainly support the production itself, as well as the collection of all preparation plants. The team is supplemented by long-term interns from all over the world.

In addition, there are personnel for the administration, the packaging and dispatch of the preparations as well as for the soil laboratory. The process of preparation making is now managed by a former member of the Domaine Saint Laurent farming community. In this respect, the regrettable ending of the cooperation with the Domaine thus represents a gain for the operation of the Massons.

View from Les Crêts into the countryside of Burgundy.

First steps in preparation practice

Pierre

Pierre comes from a strictly Catholic family and for him a priestly education was originally planned. He completed his school education with the Baccalauréat agricole in 1964 and began his work as an agricultural consultant with the main focus on animal breeding for the French Ministry of Agriculture.

In 1972, he participated in a three-day course at Château de l'Ormoy, presented by three pioneers of French biodynamic agriculture. Of the three main themes of cosmos, spirituality and closed farm organism, he was particularly attracted to the latter, since here there was opportunity for practical work, especially with the preparations.

In the meantime, he had acquired his first farm and was able to take his first steps in practical biodynamic work on the 1 ha (2½ ac) estate.

In 1976, he worked on a new 5 ha (12 ac) property, and that year, when a great drought prevailed throughout Europe, he was confronted with questions from his neighbours about why, in contrast to them, his cultures thrived so well and why his dung did not smell. This was evidently a result of the use of the preparation in the vegetable garden, in manure processing, in the compost and in cereal production.

In 1978, Christian von Wistinghausen visited France and gave a lecture at Château de l'Ormoy, where Pierre was also present. Pierre still cannot explain why he was the only one who was given the only box of preparations brought along by Christian von Wistinghausen, and through which his relationship with the biodynamic preparations was deepened.

Until then, he had not made the preparations on his own. From 1980 onwards, on a 25 ha (62 ac) farm, he started the production of the field-spray preparations and some compost preparations. He gradually converted the dry storage of the preparations to moist storage, since he always had an unpleasant feeling when using dry preparations.

In 1992, Pierre joined the seven-headed farming community of the Domaine Saint Laurent, where he was responsible, among other things, for making the preparations.

In 1994, Pierre met Alex Podolinsky on his European tour and spoke to him about the proper production and storage of the preparations. Later, visits to Agrolatina in Italy and to Carlo Noro followed. Pierre felt he had gone a step further, and his doubts gave way to growing confidence to handle the preparation work properly.

Vincent

Vincent's childhood was of course connected with the agricultural work of his father. He spent his school days at Waldorf schools in Lyon and Paris. In retrospect, Vincent often wonders why others were wondering about his views while he considered them to be completely normal. He regards this as a consequence of his school education, which he now perceived as a manifestation of a life-style.

At nine he was sent to a boarding school and only saw his parents during the summer holidays and in winter when he would help with wood chopping.

After taking the baccalauréat in Macon in 1998, Vincent spent three years on various trips and different work stays, among others, at Camphill institutions and on different biodynamic farms.

Then, influenced by the attac movement, Vincent decided to study economics and law.

In 2005, his father asked him to follow up on Les Crêts. This began a phase in which Vincent systematically learned the perfection of the preparation work of his father, in order to gradually take over the management.

In 2010, Vincent gave his first biodynamic training courses, which is now a major part of his activities. In order to make this possible, the practical production of the preparations has to be handed over to the employees step by step without losing sight of the high standards developed, which form the basis of the core activity at Les Crêts.

How the work developed

Pierre and Vincent emphasize that for them their way of working is never static. 'Today we have found a way that works well. But we're not static and fixed in our work, maybe we'll change it in ten years,' Pierre said.

Pierre and Vincent see the indications in Steiner's Agricultural Course as the basis for their own work, their knowledge, and their development. There are also ideas based on personal experiences, experiments, and the exchange with Alex Podolinsky, as well as 'a lot of knowledge from the farmers themselves.'

In the meantime, the Massons have gained a certainty in their preparation work. This makes it possible, by mere observation, to recognize the efficiency of the preparations, expressed mainly in the colour of the soil, the more straightforward and airier nature of plant growth, and the better root-growth in the soil. This can be documented by photos, especially when there are untreated reference plots. The observations are supplemented by the preparation of soil chromatography as well as soil analyses, which can now be carried out partly in their own laboratory.

However, the Massons only regard the preparation work in connection with biodynamic work as a whole, in which, according to their opinion, there are still many possibilities for further development. Many approaches, which have fallen into oblivion in the meantime, have to be resumed, among others the work of Eugen and Lilli Kolisko. For this, it is imperative to also tread non-dogmatic paths. Their security in their preparation work cannot be a reason for them to stagnate in the development of their own perception.

Pierre and Vincent Masson's understanding of the preparations

For Pierre and Vincent there is no doubt that something happens on the karmic level and in the spiritual world when working with the preparations. For the Massons, the most important aspect in understanding their preparation is the physical effectiveness of the preparations: 'Everything we do, what we are doing at the moment, works on the spiritual and karmic level. I want it to work at this level, but also to work as a farmer', Pierre says. For him, it is an important part of the preparation work to explore and understand the relationship between the preparation plants and the planets. In addition, there is a reference to the ideas of anthroposophic medicine. More specific, the compost preparations are considered by the Massons as models for what is to be stimulated by them in the soil and in the compost.

Anthroposophic medicine as a way of understanding the preparations

For Pierre, the basics of anthroposophic medicine are a way of understanding and classifying the plants, animals and animal organs selected by Rudolf Steiner. According to Pierre, the idea of anthroposophic medicine is 'to find in nature an organism or an organ, an animal or a plant, that is, something that has developed a certain type or character' and to develop a remedy from it. As an example, he refers to oysters, which filter a lot of water and build a strong protective cover (oyster shell) from the filtered lime. Calcium carbonicum (the remedy that is extracted from the lime of the oyster) is used for the strengthening of the calcium metabolism, and to give security and serenity. This idea to use the spirit and knowledge of a plant, animal or animal organ 'is also an important idea for the preparations' for Pierre. On this basis, he is able to comprehend the choice of the organs and plants that Rudolf Steiner made. 'These animals and their organs have developed a very particular character, they have something essential which corresponds to the model of the plant.'

Pierre illustrates this by the choice of yarrow and deer bladder: 'What is the genius, the spirit of the yarrow? The yarrow is the plant of regeneration, Venus, it relates the etheric and the astral field. In the body it is the kidney, which is considered an organ of regeneration – it is an organ of Venus. The kidney, however, can not be a sheath for a preparation. So you have to find another organ, which is connected with Venus – that is the bladder. The cow is not a 'donor', it is a metabolic animal and is not sufficiently associated with Venus. The choice is the stag, it is hyper sensible, that is Venus! The decision for the male animal is understandable, if one thinks of the new antler formation every year, which carries the greatest regenerative power in itself. The bladder is also the organ of sensitivity – this is shown by the fact that you have to pee when you are excited. This is really a choice Steiner made, he recognized what qualities the stag bladder unites. Each preparation is to be a model of development, a model for what it is stimulating in the earth and compost.'

Compost preparations as carriers of planetary forces

The contributions of Bernard Lievegoed are very important for the understanding of the preparations and their link with the planets and organs to the human or animal body.

In addition to the image of seeing organs as models, Pierre still gives a second image: the image of preparations as organs in the compost pile. In this respect he also describes the relation of preparations and planetary forces, 'because they are really organs that are made in the compost pile. With the yarrow is given a renal system. With chamomile it is the activity of the digestive and respiratory systems. The activity of Mercury. With the dandelion you add the activity of Jupiter. Liver. Storing substances and all that. With nettle comes the quality of Mars. With valerian you add a whole system, starting with the formation of blood in the bone marrow until the rupture of the red blood cells in the spleen. It is a global picture of processes, they are all biological processes. And we would like the compost to become an organism with organs. Preparations play the role of organs. Valerian has two roles. First of all, the role of an organ within the compost, and on the other hand a kind of protective envelope, similar to the role it had in the sky.'

For Pierre, valerian is related to the phosphorus processes: it influences the phosphorus metabolism in the soil and forms a heat envelope around the compost pile. It is governed by Saturn. According to Pierre, Saturn is the planet that separates the solar system from the region of fixed stars, creates a network between stars and planets and acts in a fine, subtle sphere.

Example of the research work of the Massons

Pierre shows a photo documentation of the research on the application of the prepared horn manure preparation (500P) on a wine-growing business. On a surface with very stony ground, even after repeated application, no effect of the preparation could be recognized. Pierre and Vincent suspected that because of the stony soil most of the preparation evaporates before it penetrates the soil. As a consequence, they recommended an increase in the amount of water and a strict application in overcast and cool weather. By observing these measures, the desired effect could be achieved.

Understanding and assessing qualities and affects

For the Massons the notions of quality and efficiency are closely linked. From their point of view, both the quality and the action of a preparation must be visible to everyone both objectively and physically – be it the farmer or the visitor to the farm.

According to the Massons, the basis of a good job dedicated to the preparations is the production and use of humid and colloidal preparations. They must, from their point of view, be the object of a strong development and transformation during their stay in the earth, and afterwards of their storage in the cellar, in order to obtain a final, humid, humic, colloidal preparation. Moisture, odour and colour are key criteria for quality.

The effectiveness of the preparations must be physically visible and show up in the soil and plants within one year and no longer.

Whether it is their own research work or their recommendations to farmers who have adopted the biodynamic approach, untreated control plots are an absolute necessity for practice. However, the aim of these plots is not 'to carry out trials with many analyses, but that the farmer can see and recognize the result, the positive change,' Pierre said.

With all the experience that Pierre brings to his understanding of the preparations as a whole, he has not lost the ability to be amazed and enthusiastic about results and stressed:

'It's already been 42 years since I started working with biodynamics. And every time I see something working I rejoice. It stimulates me again, and I am surprised every time that it works. This is just as important for people as it is for me.'

Qualities

There are three physical properties that Pierre and Vincent always check:
- Moisture. In order for a preparation to work, it must always be 'alive'. For the Massons this is only possible with moist preparations. The moisture is optimal when the preparation is completely moistened, holds water, but does not drip. All preparations in the preparation cellar of the Massons lie on the hand, without losing water. When pressed together, some water comes out. Also in the storage vessels there is no water, everything is bound in the preparation. According to their opinion, a once-dried preparation loses a good part of its effect, since without

water the microbial life dies and the preparation is taken from the life process.
- Smell. The smell of the preparations should always be 'pleasant', and never anaerobic. A light floral, ethereal fragrance can also be produced, which however has nothing in common with the plant in question. Preparations that have once dried up or smell unpleasantly, are no longer usable for the Massons.
- Color. Depending on the preparation, the colour varies in different browns (the chamomile preparation is light brownish and the oak bark preparation is slightly redder than the other preparations) and becomes darker during storage.

The Massons are also looking for ways to make quality visible not only on the physically perceptible level. They are currently investigating whether round-filter chromatography can be a method for this.

In order to achieve the desired quality, the preparations have to go through a very definite development according to the Massons. The basis for this is already laid down during the harvest of the preparation plants. For example, with the yarrow, attention is paid to the fact that only the heads are cut. If there are stalk parts, the desired development can never take place according to Pierre. The preparation should undergo a development in the soil as far as possible. This first development, however, according to the Massons, is never sufficient. It continues in the preparation cellar and is guided by the described quality controls into the right direction. Of the development process, Pierre has a very clear picture: 'It is to go towards humification. And towards a colloidal substance. This is the way for all preparations. This is our way.'

As an example, the interior of a currant is what he calls 'colloidal': a substance of enormous water-holding capacity, which is flowable and still holds together.

Preparations that have not already done well in the soil do not correspond to the ideas of the Massons. They are not stored in the cellar because, as Pierre says, 'a vinegar will not become a good wine,' Pierre says. Only after a certain time in the cellar is a preparation 'ready for use', according to the Massons. Then their condition is stable and colloidal, and the moisture regulated. In the preparation cellar of the Massons, the preparation of the current year clearly shows the structure of the plant. The preparation of the previous year is definitely converted and shows no plant structure any more. This is the goal of the Massons. They never sell preparations too fresh – only sufficiently developed preparations that have participated in the 'maturing process' in the cellar. Three to four years are mentioned as possible storage and development periods. The Massons have preparations of several years in stock to compensate for possible quality or quantity problems of a certain year.

Effects

'Just feeling is not enough – there must be objectively perceptible results of the preparation work.' This is the basic statement of the Massons. They focus on the investigation of the effect of the spray preparations. An important basis for their research is the comparison of treated and untreated plots. Therefore they advise every farmer to set up an untreated control plot next to the biodynamically cultivated field in order to be able to observe the development of the soil and the plants. Concerning the mode of action of the compost preparations, studies of the Massons have been made on valerian (507), nettle (504) and yarrow (502), individually sprayed, as well as comparative investigations with prepared and unprepared composts. However, they recommend a compost heap with, and one without, preparations to observe the development. They point to Steiner and his statement that one should simply 'do' things. The objective is not to have to carry out innumerable experiments and analyses, but that the farmer can perceive a relatively positive result in the form of a positive change. The Massons cooperate closely with many of their preparation and counselling customers for research purposes.

Study of the effect of horn manure, prepared horn manure and horn silica

To investigate the effect of 500 (or 500P), the Massons use a digging fork or a tapered soil drill to collect soil samples up to 80 cm (31 in) deep of the treated and untreated plots and compare the development of the soil. The following aspects are investigated: crumbling structure, colour, water retention, smell, root penetration.

If there are no differences in the soil within one year, the way in which the preparation is applied is examined for errors. Everything, from preparation storage to the time of application, is precisely included 'because usually someone who uses the preparation must receive objective results within one year at the latest with the occurrence of a change,' according to Pierre.

The following method applies also to the horn silica preparation: Everyone, whether a farmer or a farm visitor, must be able to observe changes. In the horn silica preparation, the Massons focus their observation on the gesture of the plant: on the composition of the annual shoot, the structure of the plant, and the shape and colouring of the leaves.

As a manufacturer and consultant, it is their aspiration to demonstrate significant effects of the spray preparations no later than one year after application. In the field of viticulture, Pierre says that 15 days after application of the preparation visible changes must occur, and the difference in quality between organic and biodynamic grapes must be tasteable: 'A few days after the application of a preparation one must be able to recognize behavioural changes of the plants, the vine. With vines it works in that way, that six rows are treated with preparations and six without and immediately you see results ... we enjoy it of course very much. We go to the people during the season, and at the vintage we try the raisins. And when they soak, we taste and see the differences between organic and biodynamics.'

Preparation practice

The Massons produce all eight classical preparations, together with the Cow Pat Pit according to Maria Thun, and the 500P (prepared horn manure) according to Alex Podolinsky.

At the Massons, the production is always done in teamwork with permanent employees, employing recurring seasonal workers from the region and international interns and guests. It is important, however, that there is always at least one experienced person. 'When the chamomile preparation is produced, the quality is related to the intensity of the stuffing process. If it is not enough, it develops badly; if it is too heavily stuffed, it will become silage. You have to keep that in mind,' Pierre says.

Observation of the constellations according to Maria Thun

In general, the Massons do not pay attention to the constellations according to Maria Thun. The 'harvesting days' of Maria Thun are not taken into account during the harvest of the preparation plants, in general it should be a beautiful, rain-free day. On full moon days, and when the moon is near the earth, the plants are not harvested since these days have 'too much water' for the blossoms. However, other preparation work can be done. In this case, Pierre more likely follows the directions given by Hartmut Spiess than by Maria Thun. On all days with nodes or lunar nodes generally no preparation work is done.

Origin and quality of the materials used

The manure used for the horn manure preparation is largely derived from the cows of the Domaine Saint Laurent. The quartz crystals are collected in the high mountains of the Alps. Except for the valerian, which is partly cultivated, partly collected wild in the surrounding area, all the plants for the compost preparations used by the Massons are cultivated on their own land or on other biodynamic farms. Careful attention is paid to the correct plant condition, especially when working with untrained people. In order for blossoms to have the same condition as possible, a plot of land is first completely cleared and then fully harvested. Fresh or dried organs are used. Pierre is explicitly against the freezing of the organs because he sees it as a 'process against life.' In his opinion, fresh organs are optimal. Dried organs are also well-suited. Before filling them, they must be made supple with good, lukewarm water or a tea from the plant used.

Field spray preparations

Horn manure preparation (500 and 500P)

During the time when the manure is being collected, a part of the cow herd is placed on a pasture, where 501 had been sprayed before. When there is too much fresh grass, the cows may be fed hay. The feeding is adjusted until Vincent is satisfied with the consistency of the manure: paste-solid, not green and without large undigested portions. The manure is picked up daily by the farmer and the team of BioDynamie Services directly from the pasture, then promptly collected by the Massons and processed further. A multi-headed team is engaged in filling and burying the horns. The horns are filled with sausage tamping machines and then buried – 45,000 on Les Crêts, 5,000 on another Demeter farm. A special focus is put on the fact that they are arranged in such a way that no water can accumulate. Several layers of horns, each separated by a thin layer of screened earth, lie one above the other, but must not touch each other.

Burying the horns.

The horns lie in several layers one on top of the other (left).
Filling the horns using the sausage tamping machines (right).

Quantities and spreading

The Massons recommend the use of 100 g (3½ oz) of horn manure or 500P per hectare (2½ ac), stirred for one hour, dissolved in 25–35 l (6½–9 ga) of lukewarm rainwater. It should be carried out at least twice a year: in the spring after the frost and in the autumn before the frost – times when the activity of the soil life is highest, according to the Massons. A late afternoon with mild temperature and overcast skies is ideal; drizzle is beneficial, but in case of rain and frost no spraying should take place.

Horn silica (501)

In selecting the right quartz for the silica preparation, Pierre chooses crystals that are not completely transparent but are said to have a 'certain translucence'. This statement is taken from the medical course in which Steiner states that 'crystals are to be collected in the mountains and have a certain light permeability.' He personally also takes care that they have a 'certain form' and uses only self-collected crystals from Europe. The Massons use different methods to grind the crystals. It is important for them that the silica is always triturated in such a way that it remains colloid when moist (see image below).

Quartz for the Horn silica preparation (501).

Grinding by means of an iron rod

First, the crystal is smashed with a hammer, and small impurities such as iron or sulphur inlets are sorted out. The next step is to crush the silica with a heavy iron rod. The remaining iron particles are removed with a magnet to obtain a pure product.

Ground quartz with water forming a colloidal structure.

Grinding

A friendly ceramist has a motor-driven mineral mill that grinds up to 150 kg (330 lbs) of silica in a three-day process to the desired fineness. The porcelain bowl of the mill is filled with water and the silica is ground to an icing-sugar-like powder.

Meanwhile, the operation of a ball mill is planned at the Massons. Two different mills are available, and the one that delivers the better results will be used in the future

The ground silica is mixed with water and filled into the horns. These are buried between Ascension and Pentecost. At the beginning of November the preparation is recovered. The recovered silica is stored in a glass vessel (preferably in a place with an eastern orientation – not too much direct light) and moved once a month for light aerating. So conserved it keeps for many years, according to the Massons. Electromagnetic radiation should be avoided. If there is no appropriate space in the house, Pierre recommends a secure space in the garden. He prefers a place in the half shade.

Quantities and spraying

The Massons recommend the use of 4 g (0–14 oz) per hectare (2½ ac), stirred for one hour in 25–35 l (6½–9 ga) of lukewarm rainwater. The period of application is mainly in spring and autumn. In their opinion the effect is optimal during the phase of the strongest growth. For example, for stocking or ear formation, as well as for already fully developed fruits before harvesting. The horn silica preparation should be sprayed directly after stirring; preferably shortly after sunrise on a clear, windy morning with plenty of dew and sunshine. Spraying in heavy rain does not make sense for the Massons, but light drizzle is very beneficial.

Compost preparations

Yarrow preparation (502)

The harvest of the yarrow begins with the first mature flower at the end of June and ends in early August. The Massons harvest both self-cultivated and wild-growing plants. The flower is harvested when it is at its peak in the early morning. Flower or Fruit days are not considered separately; but wet days, knot days, full moon, new moon, and days when the moon is near the earth (watery) are avoided. Pierre can not understand the statement of Maria Thun, that the yarrow should be harvested when the Sun is in Leo (from August 10). In his opinion, the flowers are already 'too mature' and no longer provide the best possible quality. 'When the plant has reached this stage, it can no longer be transformed in the earth. The yarrow is the plant of Venus and Venus is the goddess of beauty and regeneration. If the yarrow is harvested in August, she is already an old woman.'

The Massons harvest the plant with stalks in the morning, and in the afternoon, after the harvest, the flower heads are removed with the utmost care. These flower heads are dried on grids in the shade and then stored in paper bags. The yarrow that is harvested in one year, always serves to fill the bladders of the next year. The bladders are filled between Easter and Pentecost. Before filling, both the bladder and the flowers are soaked in a tea made of fresh yarrow. The filled bladders are hung in a fly-screened box and fastened under a roof overhang. Thus, the bladders are exposed to the forces of summer.

Before they are put into the soil, the filled stag bladders are soaked in water until the contents are completely moist. According to Pierre, it is pointless to bury a dry bladder, since no reaction can take place: 'then the preparation will never develop well.' The moistened bladders then come into a weakly fired, sufficiently porous clay container. If not all flowers are well moistened, they need the first three to four months in the soil to homogenise. If the flowers are already moist, the development and humification process starts as soon as the preparation is in the soil. If it is recovered after six months, the structure of the flower is still recognizable, but the entire preparation is already completely black-humidified.

Stuffed stag bladders, exposed to the forces of summer.

Soaking the stag bladders before burial.

Yarrow blossoms in different stages. The flowers in the middle are of the quality that the Massons use.

Chamomile preparation (503)

For Pierre, chamomile is linked to the forces of Mercury. It is for him the planet of those processes in motion that contribute to a faster development – such as the transformation of a preparation in the soil. About chamomile, he says: 'It is related to calcium metabolism and regulates processes of nitrogen, as it is indicated in the Agricultural Course.' The small intestine of the cow is put into chamomile tea before being stuffed with the flowers, whuch are themselves slightly moistened. Chamomile transforms faster than all other preparations (which Pierre attributes to the strength of Mercury) and loses much of its volume during storage in the cellar. It is important for him that the preparation does not dry out, and that it retains a moist, colloidal structure with a brown colour (the chamomile is always browner than all the other preparations) and a light floral fragrance (but not that of chamomile anymore).

Finished preparation of chamomile at the Massons: reddish brown and colloidal.

Quality of the chamomile blossoms, 'like fresh – even when dried.'

Nettle preparation (504)

The Massons only harvest the lower leaves and flower tips of the nettle but no stems, otherwise no complete transformation can take place. Harvest time takes place in the autumn as does the burial. The decision was taken by the Massons to follow the information given by Steiner as precisely as possible in this regard. The indication given in the Agricultural Course is that the preparation should be buried in the ground for a winter and a summer and not the other way around. Pierre and Vincent assume that Steiner was aware of the quality of the iron. In autumn, around Michaelmas, iron from meteors comes to earth. This is a particularly qualitative iron that is integrated into the earth organism and is also found in the nettle. Harvesting nettles in autumn is much more difficult, because there are many more different vegetation stages. Also, it is no longer possible to find nettles in the same abundance as in the spring. But this greater effort leads to a better quality, according to the Massons.

The harvested material is only slightly wilted and then stuffed into clay drain pipes – a method derived from Alex Podolinsky. If the material is already too dry, it is moistened with nettle tea. According to Pierre the nettle can regulate its moisture sufficiently well, there is a hole in the bottom of the clay pipe (through which excess liquid can trickle off). Pierre does not want to put the stinging nettle into the earth without any cover: 'it is no longer found in the field. Mice distribute everything in the ground. Clay pots are also soil – just a little bit burnt.'

Oat bark preparation (505)

For the oak bark preparation the rind or bark of still-young oaks is rubbed off with a fine grater. The filled skulls enter a pit with rainwater. Every week, the pit is filled once and flows off gradually. In this way, there are alternations. During storage, it shows a reddish coloration, and rain worms can be seen. Pierre considers organic life in the preparation as a very good sign, and an indication of quality for him. Especially with the oak bark, which contains so much tannins and calcium and therefore does not offer attractive conditions for biological activity, Pierre is pleased when he sees the worms.

The differences between the batches of the preparation from 2012 to 2014 are impressive. Although the structure of the oak bark is always clearly visible (the preparation is not completely plastic), it becomes darker and finer with every further year of storage. During storage, the preparation is kept moist – water is added until it is completely absorbed by the preparation. The oak bark preparation is enormously colloidal and can hold an abundant amount of water.

Dandelion preparation (506)

The dandelion blossoms originate from the Massons' own property and from the Domaine Saint Laurent. Generally it is not harvested before 10.00am, because not all flowers are open. Only flowers that still have a closed centre are harvested. If sufficient personnel is available, it is picked without special preparation. Working with people without experience Pierre recommends (and practices) to clear a certain area of all flowers in one day and to harvest all the flowers that have sprung up the next day.

The mesentery is used as the sheath in which the intestine is suspended. This is the part that is very close to the 'mystery of digestion'. Pierre assumes that this part is directly related to the processes of the liver and is thus associated with Jupiter. The omentum, which encloses the internal organs, is not used.

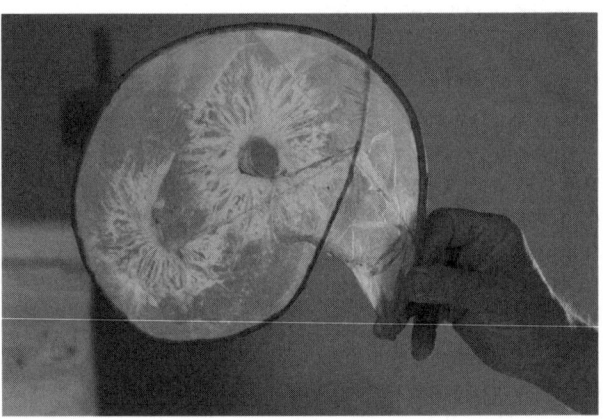

To dry the mesentery it is stretched on branches.

Valerian preparation (507)

For the production of the valerian preparation, valerian is used from their own cultivation and from wild gathering. Only the flower petals are harvested. The harvested material must not contain green parts. According to Pierre, it does not matter which day (Flowering/Fruit) or at which time it is harvested. His reasoning is that the petal is not an organ that is still under development, it has already been 'completed'. For the further processing, the Massons do not use only one method.

Valerian extract

A glass vessel is partially filled with petals and water of very good quality. The vessels are placed in the half shade for 10–15 days, then everything is filtered and filled into glass bottles. The result is a green-gold, delicately scented liquid 'with very delicate aromas.' This method has been adopted by Pierre from Podolinsky.

Pierre Masson with a bottle of valerian extract.

Applying the compost preparations

The arrangement of the individual compost preparations in the compost heap does not play a role for Pierre and does not have any influence on a better development of the compost. He describes this as pure mysticism and makes his holes 'for practical rather than for mystical reasons' with his hand 30–40 cm (12–16 in) deep.

He works with six holes in a row. There is no hole in the centre, the sixth hole is provided for the valerian preparation. For composts of 10–15 m³ (13–20 yd³), Pierre recommends the use of 2 g of preparation (preparation 502 to 506) or 5 ml (0.17 fl oz) of dynamised valerian preparation. The latter is stirred with lukewarm rainwater in a non-synthetic vessel for 10–20 mins. For compost up to 50 m³ (65 yd³), according to Pierre, 5 ml (0.17 fl oz) is sufficient for 1 l (0.25 ga) of water. For 500 m³ (654 yd³) of compost, 0.25 l (8.5 fl oz) of valerian preparation is recommended for 15–20 l (4–5 ga) of water. In addition to being placed into the hole, the valerian preparation is sprayed finely over the pile. According to Pierre, this can be done with a hand broom for small heaps, but for larger heaps he recommends a backpack sprayer with a nozzle for fine nebulization.

> *Personal perception of the preparation work*
> Wolfgang Stränz
>
> During my visit in April 2017, the recovering of the horn manure preparation was in full swing. Two workers dug out the horns, cleaned the outsides of them, and handed them one floor up to two members of staff, who carefully cleared the soil from the horn openings with spatulas. From there, the horns were handed to four employees, who knocked the horns onto a piece of wood and carefully examined the contents. Well-tested horn manure was put into a bucket; horn manure that was found to be too moist was placed to dry on a wooden board in the shade, and horn manure that had not yet been transformed sufficiently, was placed into another container.
>
> The working atmosphere seemed to be very easy going, but the team worked diligently and conscientiously, and I felt that the employees did their work with dedication. The bright sunshine enhanced the good mood.
>
> In the early morning there had been frost in many parts of the French mountainous regions, and the telephone rang ceaselessly. In the office, Vincent and his staff had to listen to lamentations from anxious customers and had to give advice on what to do.
>
> In the laboratory, the last-produced soil chromas were discussed and reflections were made on how to improve the production of the chromatography. In the mailroom, packets of preparations and dried plants were packed and made ready for dispatch.
>
> At lunchtime, the staff gathered in the dining room and everyone ate from their own lunch. There was wine, which had been sent to the Massons by the winegrowers as an enthusiastic proof of the effectiveness of the preparations.
>
> In the course of the working day, the staff received praise and encouragement from both Pierre and Vincent. Even the construction workers were honoured for their work.

Burying and storing practice

Burying practice

All compost preparations are placed into the soil in clay vessels. These are specially made by a potter at a low temperature so that they can transpire. They are bulky and evenly perforated for moisture control, breathing, and for contact with the earth. They guarantee a recovery of the preparation and protect against animals. For the containers, holes 25 cm (10 in) deep are made and they are covered with sieved topsoil or compost. The holes are marked with sticks. A distance of 10 m (33 ft) between the individual preparations is sufficient according to Pierre. At the burial site, the soil must be humus, but not humid. For the choice of the site, the condition of the soil is also important.

Storing of preparations

The finished compost preparations should not be stored for more than one week outside of the storage area. In the basement there are boxes with 10 cm (4 in) of dry peat between the preparations, and on the bottom, sides and lid of it. It is extremely important for Pierre that all preparations never come into contact with peat. Using the example of the horn manure preparation, he explains: 'The manure is on the way to becoming humus, to form colloidal structures. The peat is a model of organic structure, which regressively develops, into the direction of coal. If peat comes into contact with the preparation with its opposing evolving force, it can, for example, if it gets into the stirring barrel – reverse its effect. The 500, 500P and the Cow Pat Pit are humus models, they demand the soil to become "also such a good humus". If peat is added during storage or dynamization, the soil is proposed to become coal. And the soil now says, "On the one hand I shall become humus, on the other hand coal", it does nothing more!'

All six compost preparations are stored in glass vessels, enamelled or glazed stoneware containers. The size of the preparation container should be in good proportion to the quantity. For smaller quantities or residues of individual batches, small screw-on glasses can be used to hold the moisture. It is important to ensure that the covers are free of coatings. The cover should be only a little air-permeable – either perforated or not fully closed. At the Massons there is a plate of copper on the bottom of each preparation box. The copper plate with its laterally bent edges catches the excess water that emerges from the holes in the earthenware pots, which would otherwise moisten the wood of the crate or the peat, thereby losing its insulating property.

If a preparation is found to be too wet, it is spread out for a day or two to dry in the shade. Large, burned clay dishes are used for this purpose. Another measure used by the Massons to 'control' the 500 and the 500P and steer them in the right direction, is to insert a clay tube in the storage vessel when it is first stored in the cellar. This has two functions: first, to control the bottom of the vessel (is there too much moisture?), and secondly, ventilation of the preparation. This promotes the homogeneous development of the preparation. If the preparation is too dry, it is moistened with Easter water. The Easter water is collected from a spring on Easter morning and stored in glass balloons.

Derived preparations and other applications

Two other preparations are made by the Massons in large quantities: The Cow Pat Pit according to Maria Thun, and the prepared 500 and 500P according to Alex Podolinsky. Pierre emphasizes that the 'Cow Pat Pit' is 'not quite our thing' and is mainly produced due to the high demand. The Massons themselves let the customers decide – but recommend the 500P since according to Pierre it is 'more efficient on the soil development.'

Cow Pat Pit preparation (CPP)

The Cow Pat Pit matures in pits. These are lined with wood or bricks. For wood, poplar is used. For Pierre it is crucial to use softwood to avoid contact with resin. This principle has been taken over by Pierre from Indian models so nothing is mixed with soil. The components mentioned by Maria Thun – cow dung, compost preparations, grated egg shells and basal sand – the Massons dynamize for one hour in a concrete mixer. Then the mass is filled into the pits. In the pits the mass is regularly turned with a spade until no green parts are visible and it no longer smells. After three to six months, everything has turned into a black matter, it has undergone a humification process and is usable. It is always possible to reload the Cow Pat Pit, here the Massons follow Maria Thun. According to Pierre's view, there would be a lot of potential for improvement in the quality of the product if, as in the case of the horn manure preparation, one takes into account the processes of the year, and prepares the preparation in the autumn and exposes it to the winter forces.

500 prepared (500P)

The starting material for the production of the 500P is the finished horn manure preparation. In the next step, it lands in storage vessels in the preparation cellar. Since the Massons produce large quantities of 500P, they use 70 l (18 ga) stoneware containers in the preparation cellar. The 500 matures for three months in the storage vessel, with a drainage tube in the centre. If the 500 is too moist, it is squeezed out, if it is too dry, water is added. After three months the 500 is stabilized according to Pierre. The draining tube is then removed and the 500 is prepared. For this purpose, five holes are made in one outer circle with a small copper scoop and one in the middle (the latter is intended for the valerian) and filled with the compost preparations. The holes are then closed again. A further, three-month development follows. According to Pierre, this is equivalent to the development of a compost 'the best composts are produced after three to six months.' During this time a good microfauna and microflora develops. Pierre emphasizes that they are fully following Podolinsky's instructions for the production of the 500P and have not made any experiments to produce it. The 500P produced in this way, however, shows good results.

Summary

After more than twenty years as a practical biodynamic farmer, Pierre decided to devote his knowledge and his work to preparation making and consulting. He now looks back on more than forty years of biodynamic work and is as impressed with the effect of the preparations as he is surprised. After many experiences in biodynamic farming abroad, Vincent also deliberately chose the preparations as his field of work, and supports his father at Les Crêts as an equivalent expert. He is also responsible for the distribution of the preparations in France and some other European countries as well as the coordination of the consulting work with his BioDynamie Services company. These deliberate decisions testify to Pierre's and Vincent's expressed and palpable need for their preparation work: it is dear to them.

Their clear goal as preparation manufacturers and consultants is that the preparations should show physically perceptible effects, that soil, plants and product quality develop visibly and measurably and that the farmer experiences the practical advantages of biodynamic farming. In the view of the Massons, the application must lead to changes after one year at the latest – for example, in the case of crumbs, colour, water retention and the smell of the soil and in the habit of the plant.

To be effective at all, for the Massons the preparations must meet clearly defined criteria of quality. This includes, among other things, the fact that only moist preparations are used that have already undergone a significant transformation during recovery and are further developed in a controlled manner in the preparation cellar. The aim is a humous, stable and colloidal preparation.

For the Massons the background for their understanding of the preparation work is the Agricultural Course, a strong 'relationship' of the preparations and preparation plants to the planetary forces, and the leaning on anthroposophic medicine. For Pierre and Vincent, their study of the nature, character and 'spirit' of the plants and organs used is important. On this basis, they are experiencing, for example, the compost preparations as 'organs' which serve as a model for the correct development of compost and soil.

For every aspect of the preparation process, the Massons have precise ideas and quality criteria. There is a guiding principle here and, according to their opinion, there is little or no scope for individual adaptation of the preparation work: there is a right way for them. In the context of the interview, this is shown very clearly in the choice of the phrase 'one makes' instead of 'we make'. They always emphasize their aversion to any mystification. Within their understanding of the preparation work, they also achieve a high degree of comprehensibility through clear words – their consultancy mentality is noticeable.

The Massons work closely together with the users of their preparations. The findings of this practical research, the experience of the Massons and their passion for their work form the basis for a further development of their preparation work: 'Today we have found a way that works well. But we are not static and fixed in our work, maybe we will do it differently in ten years,' Pierre said.

Perception of the authors
Johanna Schönfelder, Kathrin Ortlieb

On the quality of the plants and the organs used, from production to storage:,the Massons work with the utmost care. When a preparation is taken out from the container the rest is carefully put back in. When working in the preparation cellar, hand washing must be carried out between the manipulations of the different preparations.

The Massons' relationship with the preparations is clear, pragmatic, joyful, enthusiastic, confident, experienced and result oriented. They have very clear quality criteria for each preparation, they certainly work with every step consciously, every detail of the process is important to them. Regarding their treatment of the preparations one feels respect for life, for each ingredient, whether it is the plant or organ or the water used. Each step of their work they can make transparent and explain it in clear terms. All mysticism is missing – they repeatedly emphasized. They work in a very precise and clean way as if with an application of medical resources. They attach great value to their work. In the discussion of how the financial value of a preparation is calculated, Vincent asserts that preparations must also be profitable. This implies for him a fair salary for his employees and a fair price for the dung and its collection work. The seasonal employees for the making of the preparations are often people who are not involved in biodynamic work. Vincent deliberately takes his time for collective breaks and interviews on biodynamic work. The Massons do their work with a lot of heart – and with a personal sense of the need for this task.

5. Antoine Fernex, Truttenhausen Farm, France

Dr Ambra Sedlmayr, Anke van Leewen

Introduction

Antoine Fernex is one of a number of biodynamic pioneers who have been active in Alsace since the early 1980s. There are some three to four preparation making groups in Alsace and around forty to fifty in the whole of France. The group around Antoine has been making preparations together for over 25 years. The core group members have a lot of experience in making the preparations and organising themselves as a group. For many years the Mouvement d'Agriculture Biodynamique (MABD) in France has been working together with this group. It is now looking to expand its sale of preparations using Truttenhausen farm as its production base.

Truttenhausen farm is located in Alsace close to the town of Barr, in mid-eastern France, near the border with Germany. It lies at the foot of the Odilienberg, a forested hill 330–370 m (1,080–1,210 ft) above sea level. France has a temperate climate and the region of Alsace is semi-continental with hot summers and cold winters (meteofrance.com). Rainfall is relatively low thanks to the rain shadow effect of the Vosges mountains. Average rainfall in Alsace amounts to 610 mm (24 in) per year (meteofrance.com).

Dr Ambra Sedlmayr and Anke van Leewen visited the farm on September 27 and 28, 2014. A guided farm tour with Antoine's wife, Gaetane Fernex, and an in-depth interview with Antoine were conducted on the 27th. The 28th was the group's autumn preparation-making day, which gave ample opportunities for observing the preparation-making process and asking detailed questions about each of them. In the evening, further information on the preparation practice was gathered during a second short interview with Antoine.

Farm portrait

Truttenhausen farm is owned by the locally well-known de Turckheim family, of whom Antoine is a grandson. Since 1981 Antoine has been renting the farm and cultivating it using biodynamic methods. The farm has 38 ha (94 ac) of land in one place and two separate plots some 8 km (5 miles) away from the main farm: a 4 ha (10 ac) plot used for vegetable production, and a 2 ha (5 ac) plot with a wheat and luzerne rotation. There is a very old and sustainably managed forest behind the main farm covering some 200 ha (495 ac). The forest provides a certain amount of shelter to the farm and good, clean spring water, but is also home to many wild boar and deer that roam around and sometimes damage the crops.

The soil at Truttenhausen farm originates from sandstone, granite and porphyry. It is fine, sandy, fairly mineralised and quite loose, and of a reddish colour. Being poorish soil, the farm is not suited to cereal production and the land is used mainly for pasture. When Antoine arrived in 1981 he began by building up a dairy herd. He wanted to keep the herd small, which meant that he had to find ways of enhancing the value of each unit of milk sold. This consideration led him to set up a milk-processing facility on the farm and undertake direct marketing. Antoine also continued to look after an orchard of standard fruit trees, 200 of which had been planted by his grandfather. The farm currently consists of pasture land, orchards, plots for field vegetables and some twenty polytunnels. There is a variety of livestock including thirty Jersey dairy cattle with their offspring and ten pigs – just enough to consume the whey and live outside on grass. There are also chickens, geese, three donkeys, dogs and cats, and colonies of bees managed by local bee-keepers.

There is a reed-bed-waste water treatment system that is used to treat the sewage from the house and small quantities of slurry produced from one section of the stable.

Most of the produce is sold directly via three markets, a box scheme with four delivery groups, and since spring 2014, a farm shop that is open on Friday afternoons. Half of the income is generated by the vegetable production, the other half by milk and dairy products. Six people work full-time on the farm assisted throughout the year by some ten to twenty volunteers and interns, and by seasonal helpers.

There is a central farmyard surrounded by the main farmhouse, the workers' accommodation and stables. The farmyard is always busy with people coming and going, including many children. In addition a public footpath passes near the yard and many people walk by.

First steps in preparation practice

Antoine is the son of a naturalist and physician who worked with tropical medicine, and of a nature conservationist. When he was a child, his father frequently took him out to observe wild animals near their home in a small village in southern Alsace. His great interest in nature motivated Antoine to study biology in Strasbourg. However, already in the first year of his studies he felt he was not learning about things he was really interested in. He then went to a small biodynamic farm to do an internship. Farming to him seemed very meaningful and so Antoine decided to skip his studies and become a farmer.

It was while working as an intern that Antoine first heard about biodynamics. The farmer there had been cured by anthroposophical doctors after an accident in the war and had then become a biodynamic farmer. He did not speak much about his biodynamic practices. On one occasion Antoine was asked to stir a preparation. It was cold and Antoine wanted to close the barn door and stir inside, but the farmer told him he had to keep the door open 'so that the cosmic forces can come in.' Antoine was struck by these words and was curious to know more. One day the consultant Xavier Florin came to the farm to give a talk and this raised big questions for Antoine. These questions led him to read the Agriculture Course and other relevant literature. Antoine was eighteen when he decided to become a biodynamic farmer. During his time as an intern, someone came and told him that Truttenhausen farm, which belonged to Antoine's relatives, would be available for him to rent within a few years.

At that time Antoine did not know of other people who were interested in biodynamics. However, in February 1981, Antoine attended the Intensive Study Weeks at the Goetheanum. There he met other people who were interested in the same questions as he was. The practical observation work done during this training and the people he met who seemed solid and down to earth, increased his trust. It 'brought biodynamics down to earth' and gave him 'the strength to carry it.' He felt ready to start work as a biodynamic farmer. Two months later he took on Truttenhausen farm.

In the early 1980s some twelve other people were starting biodynamic farms in Alsace as well. They had a great interest in working together and started a group that would meet one Sunday each month, visit biodynamic farms and train themselves.

Antoine Fernex, farmer of Truttenhausen farm.

To begin with, Antoine bought the preparations in. But he soon became interested in making the Maria Thun Cow Pat Pit preparation and then gradually all the other preparations on the farm too. The group started making the preparations together. They had already been meeting regularly to deepen their biodynamic knowledge and since at that time everybody was new to it, pooling their knowledge and resources to make the preparations seemed a very useful step to take. Nowadays Antoine is a member of the preparation-making group. He hosts the meetings and together with his colleagues from the Truttenhausen farm they are responsible for collecting dandelion flowers.

Antoine Fernex's understanding of the preparations

Antoine spoke about three ways in which he is learning to understand the biodynamic preparations. These are: asking questions, Goetheanistic plant observation, and formative forces research.

For Antoine, the preparations raise many questions, such as: Why do cows have horns? What is digestion? What is a cow's digestion? Why does one put manure into horns? Why is this buried? Why in winter? To him 'These are all helpful questions.' He lives with them and they provide stepping stones towards an understanding of the processes of nature.

Phenomenological (Goetheanistic) plant observation plays a crucial role for Antoine in developing his understanding of the preparations and processes taking place in nature. The polarities at work in the growth of a plant provide clues about its relationship to the earth and the cosmos. Plant observation work is important enough to be done regularly on the farm, and for Antoine and his interested colleagues this means getting up early in the morning. In spring 2014 they met every Tuesday from 6.00 to 8.45 am in order to study valerian, nettle and dock, using a comparative approach. Through careful and regular observation of the three plants throughout their growing season, they gained an inner understanding of, and a relationship to, these plants. They found that out of the three, the dock plant is the one most dominated by earthly forces, with its green smell and unrefined, round-shaped leaves. Valerian, by contrast, reveals cosmic influences shaping the highly differentiated leaves and giving it a special fragrance that can be experienced strongly right down into the roots. They found nettle to have an intermediate, fairly balanced, airy quality and its own scent.

Formative forces research (Bildekräfteforschung) is the most recent addition to the methodological toolkit used by the preparation-making group in Truttenhausen. There is a group of about twenty people that meet once a month in the nearby town of Colmar to do formative forces research exercises. Several of them are part of the preparation-making group and occasionally bring this approach into group meetings. Formative forces research was explained as being a form of meditative consciousness training. The first step is to find stillness in oneself and then consciously focus on the three parts of one's body and check how these are feeling. The next step is to approach the object of observation in a mood of openness, allowing it to reveal itself. After a while, one's focus is again directed towards the three parts of the body in order to observe any changes that might have occurred – in terms of movement, weight, warmth, vitality. There is then a group exchange and everybody shares what they have experienced. In this way one learns to become more attentive and find words to express these subtle experiences. In hearing what the others have observed and discovering that some of their experiences are common to all, the objective nature of some of these observations becomes apparent. With time one learns to distinguish what is a subjective impression and what is an objective reality connected to the subject of study. One becomes more sensitive to perceiving of etheric forces. Antoine and his wife have worked with one or two other people using this formative forces research technique to determine the best places in which to bury the preparations.

Making sense of the field spray preparations

Antoine feels that he has a greater understanding of the field spray preparations than the compost preparations, which still contain many unanswered questions for him. Antoine says that plants relate on the one hand to the earth and on the other to the cosmos: 'The plant's role in the universe is to connect heaven and earth. Without plants, the soil would have no connection to the cosmos. And this connection is necessary for humanity as well. Human beings cannot live without plants.'

According to Antoine, the field spray preparations strengthen this relationship in both directions: 'I believe there are two main processes working in a plant – one comes from below and one comes from above. The horn manure and horn silica preparations serve both these processes – becoming heavy and coming-into-matter on the one hand and rising up from below with the flow of water on the other.' Antoine's understanding is that the preparation that is connected to the one process (of physical manifestation or of dissolution) stimulates the opposite one: 'Horn manure stimulates the forces that come upwards from below. It helps the growth processes. Carbon dioxide is taken in and becomes matter, it densifies; plants take on form and shape and these are digested by the cow. The horns somehow bring a bit of heaven down to earth with their inner forces. All of this is then buried in the soil over winter, when the earth is really awake. This is a whole path of descent, a path from heaven to earth. And this affects the processes of growth that proceed from below upwards. The movement downward stimulates its opposite, the upward path.' By contrast, he continued, 'Silica is a mineral, a totally earthly substance. It is not like a plant that is of the air and becomes heavier. Silica is a very dense mineral and we make it very fine and bring it into the earth when the earth is most open – in the summer. This is a path from the earth towards heaven. It stimulates the forces that come from above. These are the two polarities.'

Making sense of the compost preparations

Antoine believes 'the compost preparations are more about health, digestion and balancing processes.' He felt he had a better grasp of the plants than of the animal sheaths that are used to produce the compost preparations, and gave some examples mainly concerning the oak bark preparation: 'There is an enormous power of growth in the oak tree. It comes up from the soil with enormous strength, but one can also observe contrary forces. Its movements are always heavy. It's not like the ash, something is always coming from above and pressing down against this growth. And one notices, one has the feeling, that if this counter force were not present then the oak would be an explosion with so much power coming from below. The one thing that holds these growth forces back is the bark of the oak. The growth of the oak seems to be limited from the outside. This is the way how I understand this substance ... to quietly limit the power of growth.' He added: 'I can see something similar with the skull. There, the life forces are also much quieter than in other parts of the body ... it has to be clear, calm, there is no movement up there in the brain, it fits, it's basically the same image.' He bases his understanding of the compost preparations primarily on the phenomenological study of the essential gestures of the plants and the organs.

Effects of the preparations

According to Antoine, the preparations help to intensify the life processes of the farm organism and create an outer skin or boundary that makes the farm organism healthier and more resilient. He also has the impression that the compost preparations help the composting process of the stable manure. He says the manure is less smelly when it is taken out of the stable and refers to the Agriculture Course, where Steiner describes the benefits of keeping the nitrogen in the pile.

For Antoine the effectiveness of the preparations is demonstrated by the quality of what is produced using biodynamic methods. He said: 'It's something that can be experienced every Saturday at the farmer's market, in the kitchen, and every time we eat products that are not from biodynamic production.' This experience provides the certainty that the preparations have an effect. It also shows how this effect can be experienced and followed through with understanding.

Antoine aims to use the biodynamic preparations in order to obtain high quality products in a way that is good for nature. He stated: 'The best product cannot be good if it is not good for nature.' He added: 'Nature today is totally unknown to people ... One steps on nature as if she did not exist. Beautiful landscapes with fruit trees and birds – then a bulldozer comes and flattens it all! This happens quite unconsciously in today's world. Nature is shattered to pieces to such an extent that the destruction of nature no longer stands out ... Nature is dying. I believe that when a feeling for the processes of life begins to grow as a result of the biodynamic method, a feeling for landscapes and for the whole of nature will also develop. This is because every farmer who works with this method develops a new feeling. It is also connected to our work with the preparations that directs us back to nature. For modern people, nature does not exist. Perhaps when on holidays or on TV, but not in daily life.'

Antoine went on further to explain: 'The preparations have an effect on human consciousness. Every person who engages a little with the preparations has to engage with the processes of nature. Otherwise the questions remain unanswered. And the questions are big. Why manure into a horn, why yarrow into a stag's bladder? This understanding is important because, through it, the farmer increasingly learns to work with the processes of nature.'

Assessing the quality of the preparations

The quality of the preparations is judged mainly by considering their condition when they are taken out of the ground. They should be neither too moist nor too dry and they should not have a bad smell. When asked to give his opinion as to the best preparation in their store, Antoine asked a colleague from the preparation-making group who answered, 'All are good.' But Antoine and his wife both said that a center is needed in order to study and assess the quality of biodynamic preparations and give guidance to farmers on how to improve their preparations.

The social setting

At Truttenhausen farm, preparation making is done in a group. In the early 1980s, when Antoine took on Truttenhausen, a 'wave' of people were starting biodynamic farms in Alsace. Twelve of them formed the Syndicat d'Alsace des Agriculteurs biodynamiques. They started out by learning about biodynamics together and later set up a biodynamic training course. They are still making the preparations together. When they started out, they invited more experienced people to their preparation-making day in order to show them how to do it. Xavier Florin was a consultant invited from time to time who brought a lot of knowledge and inspiration to the group. Maria Thun also came to make the preparations with them.

Asked about how the preparation work has developed, Antoine said that the group is always striving to improve the quality of the preparations and that this is the guiding motive for developing their preparation making practice. The group seeks good ideas and inspirations from various quarters. Antoine said they have no bias in taking on new ideas: 'We take the ideas from wherever they come from.'

One recent development has been a change in the way the preparations are stored. An effort is now being made to keep the preparations moist. The decision for this came about when some members of the group, who had been trained this way, convinced others that this method was better. Other group members, including Antoine, agreed to this change even though they are not entirely convinced about it. Antoine felt that the decision to store the preparations in a moist state was due to the 'power of consultants'. He went on to explain that Pierre Masson furthers the methods of Alex Podolinsky in France, and that some members of the group were convinced by Pierre Masson that preparations should be stored moist. Antoine himself believes that both ways of storing them are acceptable. He explained: 'In Australia the biodynamic method works so strongly that everybody can see it. This is not the case in Alsace, perhaps because the soil here is too fertile … the differences show themselves more strongly in a desert situation when life processes begin there. I still hear of people storing preparations in a dry state, having success with them and being satisfied. Perhaps this is not the most important question about the preparations.' For Antoine, it is the processes that take place when the plants and organs are buried in the ground that are most important, whereas what happens afterwards in storage is less so.

Since the MABD has taken on the production of preparations for sale, the organisation of the group is increasingly in the hands of Gauthier Baudoin who works for the MABD. He makes sure that all the ingredients are available on the preparation-making day and coordinates the work beforehand.

The people who are mainly responsible for the group choose the day for the autumn gathering. They phone each other to find out when they are free and when there is a suitable constellation (a Fruit day) for this work. Responsibility for assembling the ingredients is shared out on the preparation-making day. One person takes on responsibility for making each preparation and guides the group of people who choose to help with it.

Apart from the core group of those responsible, there are always more people who join the preparation-making day. It is a meeting point for those involved with biodynamic farming in the region and for newcomers to join the movement. Usually, between ten to thirty people attend.

Antoine feels there are many advantages to making the preparations in a group. The main one he mentioned is that a group provides the work with some social protection: 'When casual passers by come – and many walkers pass through the farm each day – they like to look and ask questions. With so many people being engaged doing it together, this work seems less outlandish.' He explained that the group makes the work seem more normal: 'It is after all such unusual work: taking manure in one's hand and filling a horn. We do crazy things as if they were normal.' Another comment was that working in a group helps to prevent dogmatism, since there is always a diversity of viewpoints that need to be respected.

Preparation-making group at Truttenhausen farm.

Other group members felt that it would be too complicated trying to do all the preparation work on one's own; it is easier doing it together. Making the preparations together means they also serve a social function – biodynamic practitioners come together from across the region, meet, and have an opportunity to get to know each other, share experiences and help each other out.

The Truttenhausen group makes preparations for the whole of Alsace. Everybody who needs preparations can come and collect them free of charge.

Preparation practice

All the eight classical preparations are made on Tuttenhausen farm as well as Maria Thun's Cow Pat Pit preparation. Nettle tea is used intensively in tomato production.

They don't work with the preparations on the days struck out in the Maria Thun Calendar and avoid Leaf days to prevent an excessive watery influence. Other than this 'there are practical reasons and there are cosmic reasons for what we choose to do on a given day,' Antoine explains, meaning that the Maria Thun Calendar is not strictly followed.

There is a 'holy' well at Truttenhausen farm. The water is believed to have very special qualities and it is used instead of tea to moisten most of the dried herbs prior to stuffing them into the organ materials.

Field spray preparations
Obtaining and handling horns

Obtaining horns is difficult in France, but some people have been able to establish arrangements with slaughterhouses. These people collect the horns and pass them on to the Biodynamic Association to distribute. The challenge is 'finding a person working in a slaughterhouse who has an interest in such things and who is willing to remove the horns of the older cows, but not from the bulls for example.' The group at Truttenhausen obtains their horns from the MABD. The horns are used up to ten times. Whether the horn is still good or not is tested by gently hitting it against some wood; if the sound is absorbed, the horn has already deteriorated; if it sounds hollow, it is good for further use. After taking out the 500 in spring, the horns are washed and stored in bags in the attic.

Horn manure preparation (500)

To obtain well-structured manure the cows are fed some hay while still on the pasture. The manure is collected in big, square plastic buckets and transported by tractor to the farmyard, ready for being stuffed into the horns on the groups' preparation-making day in autumn – usually a Sunday and preferably a Fruit day. Small, flat chips of wood taken from old fruit boxes are used as spatulas to fill the horns. Tree logs are used to hit the horns against as a means of ensuring that the horns are filled as close to their tips as possible. In earlier years, 300–400 horns were being filled, but since the MABD has started producing preparations for sale, 1,200–1,500 horns are being filled each year. The filled horns are placed in buckets ready for being taken by tractor to the field where they are to be buried. In the past few years, the 500 has been buried in the same spot at a site carefully selected using formative forces research. The horns are taken out in spring, after Easter on a Fruit day. Because a lot of the 500 is being produced for the biodynamic farmers of Alsace, a separate wooden box has been created to store it in. It is a classic preparation storage box made of two layers of wood filled in between with peat.

For application, the 500 is stirred in a wooden barrel using 200–250 l (52–65 ga) of water from the well. Stirring and spraying is undertaken in the evening. For big fields a plastic, tractor-mounted spraying unit is used, and, for smaller fields, copper knapsack sprayers. All fields are sprayed once a year at the beginning of March when growth starts. Sometimes a second application is done in autumn. For spraying the preparation no particular constellation is chosen, but the dates struck out in the Maria Thun Calendar are avoided.

Horn silica preparation (501)

Ground crystals are sourced from Pierre Masson of BioDynamie Services. He collects the crystals in the Alps and receives organs from the farms in exchange for ground silica. The 501 is made in spring on the same day that the other preparations are taken out of the ground. A day shortly after Easter is chosen because at that time there are more upward streaming forces at work in nature. According to Antoine this period is 'like the beginning of the year.' A Fruit day is preferred and Leaf days are avoided for this work.

The quartz powder is mixed with some water from the well and filled into the horns. After a few minutes, the mixture in the horns is like gypsum. The horns are then buried some 20 cm (8 in) deep in the ground on a special site that lies on a ley line connected to an old abbey. This place had been found and selected using formative forces research. About ten horns are filled with silica each year. The horns are taken out in the autumn when the other preparations are being made and buried. The horns are first cleaned on the outside and on the open end to prevent any soil getting mixed into the preparation. The horns are emptied into china dishes using long knives. The preparation is then stored in jars stood on a south facing window sill so that it can receive light every day.

Half a teaspoon of 501 is used for a barrel with 250 l (65 ga) of water. Stirring and spraying is carried out in the morning. Different crops are sprayed on different days depending on the constellations thought to be most suitable. Pumpkins and tomatoes, for example, are sprayed on Fruit days, grass on Leaf days and potatoes on Root days. The main crops receive at least one, but ideally three, silica sprays.

Stirring and applying the spray preparations

Antoine believes that the preparations should be stirred by hand and not by a machine, since a machine would interfere with the healthy flow of the life processes that the preparations are intended to support. The formative forces research group at the farm investigated different ways of stirring. They compared the stirring done by a machine developed in France to that of a flow form and a human being. It was observed that much less interest was given to a machine or a flow form in action compared with the stirring undertaken by a person. The stirring machine and the flow form gave out a feeling of heaviness. The flow form had a clearly horizontal movement, rather than a vertical one, which is thought to be more in keeping with the intention of connecting heaven and earth.

Reflecting on the stirring of the 501, Antoine said: 'Stirring takes one hour. When does one have so much time to consider a life process? When can I take that much time for the preparations? We have one hour to carry out one type of movement, it's not complicated. Then the questions arise. I get in touch with the surroundings. In spring I have swallows. I was there before them, but at some point I hear the first one. Then it comes out, and soon all are flying around. I hear the last owl and the sun then really rises. This is the beginning of the day when all plants have been waiting for the sun in this state before sunrise. There is an enormous tension in the plant world. And it is during this time that I have been stirring this sun preparation – in water. This means making it ever more sensitive – opening up and closing, opening up and closing, for a whole hour – the experiences of the whole hour enter into the water.'

With regards to stirring the 500, Antoine commented, 'the mood with horn manure is different, it is in the evening when dusk begins to fall. Out there spraying, a deeper connection is established with the earth. The feeling is more inward, but it is also a connection with the forces of fertility in the earth.'

When Antoine concentrates on the stirring, he uses the time to 'follow up some good thoughts, fruitful thoughts ... The good ideas about preparations normally come while stirring. Ideas that seem to be connected to reality, to truth ... this brings power and joy for both preparations.'

Asked about his experience of time when stirring, Antoine replied: 'An hour is always an hour ... we do it all by hand. And after all, it still is work ... I do sometimes look at my watch.' The barrel takes 200–250 l (52–65 ga) and Antoine says: 'It warms you, but I do still savour doing it this way.'

Compost preparations

Yarrow preparation (502)

Stag bladders are bought from local hunters. Yarrow is picked on the farm, dried and stored in paper bags until it is ready to be put in the bladders the following year, soon after Easter. To smooth out the bladders for filling, they are moistened in yarrow tea. Sometimes the juice of yarrow leaves is used to moisten the dry flowers. The stuffed bladders are then hung up outside of a window that faces the farmyard. In autumn, on the preparation-making day, the bladders are taken down and moistened in a bath of yarrow tea, before being buried in soil inside clay pots that are then buried.

Chamomile preparation (503)

Intestines are obtained from a biodynamic cow that is slaughtered in a local slaughterhouse. The intestines are filled with air and left to dry. Dried chamomile flowers are provided by the local Weleda gardens and stored in paper bags. The preparation is assembled in autumn, around Michaelmas, on the group's preparation-making day. The intestines are placed in chamomile tea made with spring water from Truttenhausen farm. The chamomile is moistened slightly with spring water and then stuffed into the intestines using various types of funnel. Many of the 'funnels' were made of PET bottles cut slightly above the bottleneck. The 'sausages' were stuffed well but not too tightly. Special care was taken to produce sausages that would easily lie horizontally in the ground.

Preparing a piece of intestine for stuffing chamomile.

Personal experience of a preparation-making day at Truttenhausen farm
Ambra Sedlmayr

The preparation-making day took place on a Sunday. This year it had the following programme: 9.00 am: Start with eurythmy with Gaetane Fernex in the barn to 'make people more sensitive to the preparations' and to help bring the group together. 10.00 am: Presentation by a biodynamic vintner of his experiences with EM (effective microorganisms) and compost teas. To help participants deepen their understanding of the preparations, a speaker is normally invited to give a contribution in the morning. The atmosphere in the barn was intense and positive. It was palpable that here people had gathered in order to find solutions for maintaining the Earth's fertility. This could be felt, even though the content of the talk was met with some scepticism too. When the time was up, Martin Quantin from the MABD ended the discussions and introduced the practical work of making preparations. Work then started on filling the cow horns. There was a festive atmosphere, with much talking going on.

I felt that this morning session of preparing together for the day's work and making ourselves more receptive to the preparations, brought the intention of doing something good for the Earth into consciousness. It also brought the group together.

After stuffing most of the cow horns, there was a bring-and-share lunch on a meadow near the farmyard. Farmers brought food from their farms and the vine growers brought some wine and grape juice to drink. There was a social mood, everybody spoke to, and was interested in, one another and was always ready to get to know someone new. It was also very relaxed, with some people just lying in the meadow and enjoying the sunshine.

In the afternoon, the activity resumed in the yard. There were six stalls where the different preparations were being assembled. Everybody was very focused and worked quite fast, especially those with the most experience. They worked together in pairs or small groups in the most efficient way (for example, when stuffing dandelion or oak bark). Later, as the sun was almost setting, the preparations were buried. Now there was a mood of gratitude and satisfaction.

Nettle preparation (504)

The nettle preparation for the whole group is often made on Truttenhausen farm. A scythe is used to cut the aerial part of the nettle plants in May or June. They are then left to wilt before ripping off the leaves. These are placed in a box made out of tiles that are placed together directly into the ground and surrounded by peat. The nettle stays in the ground for one year. The preparation can easily be found when it is dug up, since it is surrounded by tiles.

Oak bark preparation (505)

Since there are restrictions in the European Union on the handling of cow organs as a result of the BSE crisis, a horse's skull is used for making oak bark preparation. The skull is placed in the compost heap for about three months, so that only the bones are left. Each skull is used only once since it is broken to take the preparation out. The oak is grated directly from a tree growing on a biodynamic farm using a cheese grater. This produces a fine, dry substance that can be filled into the skull using a funnel. The opening of the skull is then closed with bones, covered in clay and finally tied together using a string. The stuffed skull is then placed in a barrel below a downpipe. The skull is covered with mud. There is a hole in the middle of the barrel, so that the skull is not always covered in water, but depending on the weather it can be aerated or even dry out.

Dandelion preparation (506)

The mesentery is obtained in the same way as the intestines. The mesentery is stretched and left to dry on fresh thin branches in a shape that later makes it easy to form pockets when assembling it. At Truttenhausen, dandelions are harvested from the fields. They choose flowers that are just beginning to open so that they achieve a high-quality, dried herb, with few or no flowers that have gone to seed.

The dried mesentery is placed in spring water together with some dry dandelion flowers, so that it becomes supple again. Two people work together to make the preparation: one person holds the mesentery open and the other person stuffs the dandelion (moistened with spring water only) into the purse. They take care to make rounded forms, as their understanding is that the sphere shape has special qualities and that these help to mature the dandelion preparation. The round packets are tied together using a cotton string to ensure that they hold together.

Dandelion preparation is the one that Antoine has studied most. Goetheanistic observation has made it clear to him how open the dandelion plant is to receiving light: already the leaves, so close to the ground, show this tendency of liberating themselves from the subterranean forces and opening themselves to receive light.

Dried peritoneum viscerale spanned on supple branches.

Valerian preparation (507)

The valerian preparation is produced in a nearby garden by the Weleda company who have the equipment needed to juice the valerian flowers. One spoon of valerian flower juice is used for 5 l (1½ ga) of water to spray the manure heap. If, however, there is a pile of manure that has never received the compost preparations, 5 full spoons are used.

Applying the compost preparations

Antoine has developed his own way of applying the compost preparations to the stable manure. He has a special box with small amounts of the six compost preparations standing ready in the cow shed. In winter, when the cows are inside, the first thing he does in the morning is to apply one compost preparation. Each day he applies the compost preparation that corresponds to the planet belonging to that particular day of the week. He associates Monday with oak bark, Tuesday with nettle, Wednesday with chamomile, Thursday with dandelion, Friday with yarrow, and Saturday with valerian. He takes a small amount of the chosen preparation and places it on top of cow pats. This is done because Antoine feels the cow pats are the areas where the life processes are most active and where the preparations can be of most help. In the process of applying the preparations he tries to connect to the planetary qualities of the preparation, while at the same time seeing it is an opportunity to check the health of the animals by observing the individual cow pats. When the manure is taken out of the stable and put together in a heap for composting, it is prepared once more, this time in the classical way.

The gardener also makes some compost for growing seedlings. The preparations are applied every time a new fermentation process begins. Five holes are made in the heap and a small amount of preparation is placed inside, unless the materials to be composted are very brittle, in which case some mature compost is used to make balls in which to contain the preparations. These are then put into five holes in the new compost pile, which is then sprayed with valerian preparation.

Burying and storing practice

Burying practice

For the last two or three years the preparations have been buried at specific spots on the farm that have been selected using a combination of convenience (a field close to the farmyard) and formative forces research (to find ley lines). For the burying of the 501 preparation, a dowser found a ley line that he recommended. To protect the stuffed organs from mice while they are in the soil, they are placed inside clay pots. Inside the pots, the stuffed organs are surrounded by soil that has been enriched with compost (to turn the poor soil of the farm into 'living soil'). These clay pots are then buried in selected places on a pasture, where the soil is not very rich in humus. Four poles are used to mark the corners of the spot where a preparation is buried. Detailed notes are taken on how much of the preparations are buried at each place, so that all of it can be retrieved.

Storing the preparations

For storing the preparations there are two wooden boxes that are kept at the entrance of a storage area, near the stable and the farm-yard. One box is used for storing the 500 only, as there are always big quantities produced. The other box contains the compost preparations. The wooden boxes have two layers of wood and, in between them, a layer of peat. Each preparation is placed in a separate glazed clay pot.

Derived preparations and other applications
Cow Pat Pit preparation (CPP)
The Cow Pat Pit preparation is used whenever composting processes in the field are to be encouraged, for example when the leaf remnants are harrowed into the soil after harvesting pumpkins.

Horsetail tea
A decoction of horsetail is made regularly, every week, to spray the tomatoes in the twenty polytunnels, with the aim of preventing fungal diseases. Nettle and horsetail are placed in cold water and heated to the point of boiling. This decoction is then diluted with two-thirds the amount of water. At the end, a small amount of oak bark preparation is added.

Summary

Antoine has a questioning mind, and the preparations challenge him to feel his way into an understanding of the processes of nature. The care and interest for nature are at the heart of Antoine's work.

The main tool used by Antoine to develop his work with the preparations is Goetheanistic plant observation, which he learned about during the Intensive Study Weeks at the Goetheanum in Dornach, which he attended in the winter of 1981. This method still informs much of his study of the preparations even though it is now complemented with formative forces research.

A speciality developed by Antoine is his way of applying the compost preparations in the stable manure. He applies one preparation every day, choosing the preparation corresponding to the planet of the respective weekday.

Social support was important for Antoine from the outset. It was the social support he found in Dornach during the Intensive Study Weeks that encouraged him to start biodynamic work himself. Later, the group of biodynamic practitioners that met regularly played an important part in Antoine's journey, and they are to this day providing the social base within which his preparation practice is embedded.

Within the preparation-making group responsibilities are clearly but flexibly shared. The fact that the preparations are made as part of a group means that Antoine does not have control over all the details and decisions of preparation making. But Antoine's ideas are not fixed, and he is ready to try out new things and accept practices that other group members feel strongly about. Antoine's individual focus lies more in developing his understanding of how the preparations work within the processes of nature, than in developing detailed practices. This focus allows for tolerance and productive group work.

6. Harald and Sonja Speer, Uppmälby Farm, Sweden

Johanna Schönfelder, Dr Maja Kolar

Introduction

There are currently fifteen Demeter certified farms in Sweden with a total of 941 ha (2,325 ac) under biodynamic management. Five of them are dairy farms and the rest focus on vegetable production.

The village of Järna (50 km, 31 miles south west of Stockholm) has played, and continues to play, an important role in the development of biodynamic work. It was here, in 1935, in the context of a curative home, that the first biodynamic crops were grown. A farm in the same village soon followed. Therapeutic education in Järna continued to expand and other anthroposophically inspired initiatives followed. The Biodynamic Association, a biodynamic research institute, a biodynamic training college, and a wholesale enterprise, as well as a dairy, are also located there.

One of the Swedish Demeter farms is Uppmälby. The farm has been under the biodynamic management of Harald and Sonja Speer (born 1931 and 1934 respectively) since 1974. Uppmälby is a mixed farm of 10 ha (25 ac) with vegetable production and sheep being the main activities. Despite its (small) size, Uppmälby is a full time enterprise that provides a secure living for Harald and Sonja. Their farm also provides everything that is needed for making the preparations. All the plants and the quartz needed are found in their own fields, as well as their speciality; the manure and organs from their own sheep.

Uppmälby is located in the Sörmland region of eastern Sweden not far from Järna. Sörmland is part of the mid-Swedish basin. The region as a whole is very fertile and intensively farmed. At the same time woods, lakes and gentle hills make up the landscape, and the local primary rock (gneiss) is often visible on the surface. It is a landscape in which cultivated and natural areas continuously alternate with one another.

In contrast to the more northerly parts of Sweden, the climate here is still warm and temperate, characterised by late but warm summers and mild winters. This means that although average temperatures during the warmer months are above 20°C (68°F), the annual average remains between 6–8°C (43–46°F). Uppmälby lies 30 m (100 ft) above sea level and is about 56 km (35 miles) from the Baltic coast. Annual rainfall averages 400–500 mm (16–20 in) with the region typically having less precipitation in early summer. Because they have a right to withdraw water from the lake bordering on their land, the farmers are able to withstand periods of drought at Uppmälby. The farm's soil is a light loam with a pH of 5–6.

Johanna Schönfelder and Dr Maja Kolar visited Uppmälby on April 7 and 8, 2015. The farm tour, as well as the in-depth interview, took place with both Harald and Sonja together. Because the preparations on Uppmälby are only dug out at the end of April or beginning of May, they could only look at those made the previous year (in 2014) as part of the discussions around preparation making practice. The researchers did not participate in preparation work.

Farm portrait

During the 1970s the official policy of Sweden was to dissolve small farms like Uppmälby and incorporate them within large farming operations. Harald and Sonja came across Uppmälby through a newspaper advert and took over the farm in 1974. The land belonging to it was then being rented by neighbours and the farmhouse used for holiday accommodation. Thanks to the closure of small farms in the vicinity, Harald and Sonja were able to inherit farm machinery suited to a farm the size of Uppmälby.

Harald and Sonja had made it clear from the beginning that they wanted to farm biodynamically and make a living from the farm. Neither the authorities nor their neighbours were able to understand this and were of the opinion that 'it is not possible to live from a farm of that size.' From the beginning, Sonja and Harald kept a diary of their progress. Now, looking back, Harald and Sonja are amazed at how 'effective one can be in the pioneer phase.' The farm was, however, only able to develop 'very slowly'. Investments could only be made when the money was available and not through loans. They found nothing wrong about this slow progress and in fact felt that it was 'just right'.

Harald and Sonja turned Uppmälby into a mixed farm. The total area of 10 ha (25 ac) was divided up into 1 ha (2½ ac) vegetables, 6 ha (15 ac) arable, 2 ha (5 ac) permanent grassland and 1 ha (2½ ac) woodland.

The livestock included 21 sheep (horned Gute sheep and hornless Leicesters), 30–40 chickens and between 3–20 ducks. The sale of vegetables generates most of the farm's income. Some forty vegetable boxes are delivered to individual households, restaurants, kindergartens, churches and health food shops. This means Harald drives his car more than 500 km (311 miles) each week. The farm has altogether around one hundred regular customers. Harald and Sonja are fully engaged on the farm and are often supported by students doing their practicals.

Uppmälby dwellings and herb garden.

Voluntary Simplicity

Uppmälby is characterised by the large amount of hand work, minimal technology and limited specialisation. Harald identifies strongly with a Swedish initiative that calls itself 'Voluntary Simplicity'. Saving energy and making the ecological footprint as small as possible 'also makes economic sense – if one is satisfied with less, there is no need to exploit animals and plants so much.' Of his farm he said: 'We have a lot of things here which, on their own, would make no economic sense. Together, however, they form a viable and satisfying whole. One could say, "Hens are not profitable, get rid of them," or, "Sheep are not profitable, get rid of them." I would like to describe our farm as the opposite of being specialised. And that is always very worthwhile. "Voluntary Simplicity" should never be equated with poverty. We can afford everything we need.'

First steps in preparation practice

Harald was born in Silesia (most of Silesia is today in Poland) and grew up on a farm. He was deported to West Germany after the Second World War. There he trained as a garden designer. Having spent most of his youth 'imprisoned' by the war, he took the first opportunity to travel abroad and, through connections with a friend, arrived in Sweden. He deepened his training further by studying garden architecture. Harald worked for twelve years as a garden designer. He didn't gain much satisfaction from the mountains of paper work and was very disappointed with the contrast between the 'dream design' and its implementation in the garden. Sonja grew up in Stockholm, studied art and illustrative design and worked for a long time in commercial advertising. When the architect's office where Harald was working, received a request to design the buildings for nuclear power stations, they both knew the time had come for them to give up this work and see about finding their own farm.

Through their beekeeping in their garden in Stockholm, Harald and Sonja got to know two of Sweden's biodynamic pioneers – Alexis and Emmy Blomberg. Once they had found Uppmälby, Harald followed the Blombergs' recommendation and went to Järna to take part in the so-called winter course – an 'Introductory Course in Biodynamic Agriculture'. He said that when he started: 'I was very familiar with agriculture but the biodynamic approach was completely new to me.'

He was also introduced to the preparations during this winter course. Harald mentioned Bo Petterson and Kjell Arman as important teachers. Petterson was an agronomist researching the effects of the preparations on food quality and soil development who sought, through his research, to make biodynamics accessible to conventional farmers. 'It was important to me that a scientific approach had been there from the beginning,' said Harald. Arman was a trained pharmacist and co-founder of the biodynamic work in Sweden. His ability to 'make biodynamics accessible' was inspiring to Harald. 'For me, when I heard about some things for the first time I said to myself: I cannot grasp this, I can't take it on. But then ... I had such trust in these people. They had something important to say, something I needed to hear a second time, or perhaps from someone else, from another angle or the view of a third person. In this way, gradually, a conviction grew in me. This was my path. There was no revelation that suddenly made me into an anthroposophist who was able to understand everything. I would never declare that – even today. Having a positive sense of being on the right track, however, that is what was important.'

Harald and Sonja in the greenhouse.

In the beginning, because they had to save money for the farm, Harald drove to Järna on his own. When he was asked during the course where his wife was he knew that it 'is very important to develop in tandem.'

Harald said, 'Working together and sharing the same objectives has been crucial to making this project work, despite all the doubts that we were surrounded by. It can work if we share the same aims. It is also important for the male and female elements to balance one another out.'

The images presented by Kjell Arman opened the door to biodynamics for both of them. He explained, for example, that the concept 'biodynamic' is made up of 'bio', which means life, and 'dynamic', which stands for force. But even more important than the best explanations was finding 'people to have faith in.'

Despite their age, Sonja and Harald Speer are not actively seeking someone to take over their farm. They trust that there will be a solution when it is needed. An intern has shown interest in taking the farm on; he will spend the summer of 2016 on Uppmälby.

How the work developed

When they started working on their own farm, Harald and Sonja bought the preparations from the Biodynamic Association in Järna. As well as attending the winter course, they learnt how to make the preparations from

Wilbert Beyer, a biodynamic farmer and veterinary practitioner on whose farm a group of biodynamic farmers and gardeners met in order to make the preparations. They also learnt from Thomas Lüthi at Skillebyholm in Järna. Just as with the farm, work with the preparations developed slowly but steadily, from buying them in to making each one of them themselves. 'It is important for me that it grows slowly and organically – in all areas', explained Harald. 500 and 501 were the first preparations they started making.

The decision to make all of their own preparations arose out of a conversation they had with Maria Thun. Her thoughts about how important it is for the preparations to be made on the farm if at all possible had 'somehow convinced' Harald. She said that 'just like animal feed, the preparations should also be produced on the farm.' Added to this were the problems connected with the BSE crisis – the abattoirs were no longer able to provide the organs. Small animals, however, could still be slaughtered for home use and so the idea arose of using the organs from their own sheep. It was especially the need for a skull when making oak bark preparation that led to the decision to work with sheep. Maria Thun recommended that 'a home produced sheep's head is always better than a bought in cow's head.' The decision to use sheep manure came about because Harald was unable to find high quality and well-structured cow manure in the region. He was keen to follow the emphasis he felt Steiner had placed on using 'well-formed manure' – something that he found his sheep were able to provide in a wonderful way.

Harald and Sonja Speer's understanding of the preparations

The effectiveness of the preparations is visible only in the long term.

As a farmer, Harald sees no quick way of determining whether and how a preparation works or what influences the constellations bring to bear. Referring to the Maria Thun Calendar he says, 'it can depend on so many things.' They both declare that they cannot perceive any immediate effects, nor do they have any supersensible perception of them. They believe that to make an effect objectively visible, comparative trials would be needed with treated and untreated plots – 'for which sadly we have no time'. When Harald applies the spray preparations he says quite candidly: 'I really do not see a difference after I have sprayed.' For him there is only a 'difference in feeling.' He has a feeling, for example, that the spraying was carried out at the right moment, soil moisture was optimal and that spraying the preparations then 'felt right'.

Over the long term, however, changes do occur and Harald sees these as being caused by the preparations: the clay soil has become lighter and better able to grow vegetables, plant diversity has increased and there is a wider range of birds living on the farm.

The preparations work together

Sonja and Harald believe that each preparation 'acts in a quite specific way', but that any visible effect is the result of using all the preparations. With regard to the farm's soil, the ability of which to grow vegetables has been steadily improving since converting to biodynamics, Harald explained: 'I am not able to say whether it was due to horn manure, compost preparations or the lucerne, it is all of them. Kjell Arman, whom I mentioned before, had a beautiful image. He said: 'There are many different things to look out for. If one thing is absent the effect is not so noticeable, but if many things are not there then we suddenly find that it is no longer biodynamic. It is a cumulative effect. We biodynamic practitioners have a wealth of possibilities for helping nature.'

Harald and Sonja are convinced that the preparations are 'one of the many things' making Uppmälby what it is. In the compost, for example, Harald experiences the collective working of the preparations like this: 'The compost is like a cow lying there and digesting – its organs are the preparations. There is always something devotional about building a compost heap. The organs inside it – like lungs, heart, liver, kidneys – work together, and the one is no more important than any other. If the preparations weren't added to the compost their guiding, participative and directing activity would be missing.'

Working with the preparations: from duty to necessity

Working with the preparations is no chore for Harald and Sonja. To them, it is important to be fully present in the work. At the beginning, Harald applied the preparations because they belonged to the biodynamic approach. It is now a matter of course for him and a necessity. It constitutes a significant aspect of his work on the farm and he enjoys doing it.

The social setting

Harald is convinced that both he and his wife would have been 'finished' long ago if they had continued working in the city. At over 80 years of age they are still running their farm and believe strongly that the way they are doing it is good for body, soul and spirit. Part of this 'way' is of course also the communal work with the preparations. When preparing the compost heaps at Uppmälby, interested people are always invited to join in. Participation allows Harald to enter into conversation with them about biodynamic agriculture. He meets people where they are. He introduces complete newcomers using the image of 'the compost heap as a cow'. With an engineer he speaks about the preparations as catalysts of the composting process, while a school child is given yet another picture.

He is continually amazed at the drawing power of the farm and can only think that it has to do with the preparations and their effects. 'It cannot be due to the few old machines we have, it must be something else,' he reflected. They often have so many visitors that there is hardly any time left for work. But 'it is hard to tell' what is actually drawing them.

Another source of strength for biodynamic work come from engaging with something beyond professional interests.

As a newcomer to biodynamic work during the winter course in Järna, Harald experienced something that has inspired him ever since. Many practical aspects of biodynamic work were discussed during the course. One of the teachers – Arne Klingborg, a co-founder of the Järna initiative – introduced an evening activity that had no direct connection to the professional studies: he led conversations about works of art. They had 'no practical value' and were of 'purely human' interest. It was particularly while Sonja and Harald were building up their farm that this principle proved a real help. To this day they go out regularly to visit an art exhibition or see a play; music is also very important to them.

Preparation practice

The preparations are made on Uppmälby, often together with trainees. The plants needed are all found growing wild and are harvested from the farm. The preparations are mostly made using the organs and manure from their own sheep; if they are insufficient they are supplemented with organs and manure brought in from cows on a friendly biodynamic farm. The stag's bladder is supplied by a local hunter. Whenever possible work with the preparations is carried out according to the Maria Thun Calendar. All the preparations are put into the earth around Michaelmas and are taken out at the end of April or beginning of May (apart from the 501 and stinging nettle preparations). This has practical reasons: until then the new preparations are not needed since there are leftovers from the previous year. Also, the soil on Uppmälby begins to dry off at that time. Harald and Sonja always produce more preparations than are needed for the coming year. Their reasoning is that 'things can go wrong sometimes and it is better to have a bit in excess, then we can also give preparations away for our customers to use in their own gardens.'

Field spray preparations
Horn manure (500)

Manure from Harald and Sonja's own sheep is used to make the 500. Because the available sheep horns are not enough to produce sufficient 500 for the whole farm, a few cow horns are also bought in and filled with sheep manure. The manure is collected from the field where the sheep are grazing as soon as possible after it has been deposited so that there is not too much earth mixed in with it. The manure is put into the horns the next day. The horns are placed upright in the ground with their openings downwards because 'they should ray back the astrality – I find it best to arrange them in the ground as they were arranged on the head of the sheep or cow', Harald explains. About twelve horns are filled each year around Michaelmas. They are buried in deep, well-drained soil. The pit is about 40 cm (16 in) deep so that the horns are in the topsoil. The 500 always turns out well at Uppmälby. Harald has never had the problem of producing poor quality horn manure. The 500 is stored in a clay pot surrounded by a layer of peat in a plastic bucket and covered with a peat cushion (a plastic sack filled with peat).

In spring, the pastures and arable land is sprayed once and the vegetable land twice with horn manure. The fields down to autumn sown grain, receive one and often two sprayings. Spraying is carried out in the afternoon or evening when it is misty or when there is a lot of moisture in the air. A portion of the 500 the 'size of a hen's egg' is stirred in 15–20 l (4–5 ga) of water. Harald does not know exactly how many grams or ounces it is. Five portions of the 500 in 125 l (32 ga) water is needed to spray all the fields once. It is stirred by hand for one hour. Valerian is sometimes added in the last 15 mins of stirring. According to Harald 'the warming effect of valerian helps to protect plants against frost.'

Storage of the 500 at Uppmälby.

Horn silica (501)

Quartz is found as a constituent of the local rock. It is broken up with a hammer and subsequently ground in a mortar. The end result is not entirely flour-like but contains rougher particles too. This follows the recommendations of Maria Thun, who felt that it is not good to break down the crystalline structure entirely because otherwise the 'warmth and light quality' of the crystal could be lost. This made complete sense to Harald. The ground silica is made into a paste with water and placed into the horns. At the end of May or early June they are buried in a pit with their pointed ends uppermost. When the preparation is taken out of the earth in autumn, it is slightly moist and is left to dry. The finished preparation is then put into a glass jar and stored on a sunny window sill for up to three years. Before use, the 501 is also stirred for one hour – one portion (about 1 tsp) in 15–20 l (4–5 ga) of water. It is sprayed out early in the morning while there is still dew around. Every crop is sprayed once and often twice. Sometimes during the last 15 mins of stirring, horsetail tea is added in order to help prevent fungal attack.

Stirring and applying the spray preparations

The field spray preparations are stirred for one hour by hand using a birch broom in a wooden barrel (125 l, 32 ga maximum). Rain water or water from the lake is used. It is sprayed with a knapsack sprayer using different nozzle sizes – large droplets for horn manure, fine mist for the 501. Harald and Sonja appreciate stirring by hand. It allows them to enter into the process with feeling while at the same time observing themselves and the process of stirring in 'a devotional mood'. Harald has no problem stirring for an hour. He describes it as a meditation and says he feels 'much fresher' afterwards. 'By remaining truly present in feeling and in the body, and also thinking about the purpose of the preparation, these qualities also find entry through the stirring.'

Compost preparations

Yarrow preparation (502)

The last preparation plant to be harvested during the year is the yarrow. It is harvested at the end of July. Whenever possible Flower days are chosen. Care is taken to pick only the flower heads, although if some stalks come with it they are tolerated. The young, freshly opened flowers are collected and dried in a drying cabinet. Before use, both the flowers and organs are moistened with yarrow tea. The bladders are then hung up under the eaves of the house – protected from birds with wire mesh – until ready for burying in the autumn. Two or three bladders are filled each year.

Chamomile preparation (503)

Chamomile flowers are picked as far as possible without stalks and usually around midsummer. Young, freshly opened flowers are picked and put into the drying cabinet. Sheep intestines from the farm are used as sheaths. They are rinsed out with water immediately after slaughter and then frozen. If insufficient sheep intestines are available, they are supplemented with cow intestines. These are usually dried. Both the flowers and the intestines are moistened with chamomile tea before use. Sheep intestines are so small that they must be filled using a pencil. Five or 60 cm (24 in) long pieces of sheep intestine are needed to produce sufficient chamomile preparation for a year for the farm.

Nettle preparation (504)

At the beginning of July when nettles begin to flower, the whole plants are harvested. They are immediately put into a sack – usually a broad-meshed, plastic sack such as are used for potatoes. The sack serves to ensure that the nettles don't get mixed up with soil. The sack is then buried directly in the soil and dug up the following July. One sack full of green matter provides Uppmälby with plenty of preparation.

Oak bark preparation (505)

Oak bark is collected from trees on the farm. A rasp is used to gather the bark taking care to ensure that no cambium comes with it; only the 'bark that has already died' should be used. Harald selects young branches with a diameter of 15–20 cm (6–8 in). It is harvested shortly before Michaelmas. The freshly grated oak bark is then put directly into the sheep skull. Three skulls are used each year. It is explained that, 'Sheep skulls are very small but they fit well in the context of our small farm.' The bark is filled into the skull using a pencil, which is then closed with a piece of bone. Much time was spent looking for the right place for the oak bark preparation. It was decided in the end to bury the skulls in the muddy bottom of a continually trickling stream that runs through the farm. When taken out, the preparation has no smell but the structure of the oak bark is still recognisable.

Dandelion preparation (506)

Dandelion starts flowering at the end of May at Uppmälby, which is when the freshly opened flowers are picked. This means picking from the same area every day. The flowers are dried as quickly as possible in the drying cabinet. The greater omentum is used as a sheath. Both organ and blossoms are moistened with dandelion tea before being used. Three packages of dandelion preparation are needed for the farm. A piece of the greater omentum measuring roughly 40 × 40 cm (16 × 16 in) is sown together on two sides, filled with flowers and then closed by sowing up the last side. Small packets are thus created. Harald and Sonja can see no great differences between a sheep's or a cow's greater omentum: 'that from a sheep can also be very fatty and there is no difference to be seen in the finished preparation.' The dandelion always turns well, while the original structure of the dandelion flowers can still be recognised.

Valerian preparation (507)

Harald and Sonja once sowed V*aleriana officinalis* and *Valeriana sambucifolia* on their farm, and since then it has become well established and seeds itself. The flowers are gathered on a Flower day in July on a sunny morning. The flowers are pressed, although this only produces a small amount of highly concentrated valerian juice. The same amount of water is then added to the pressings as was removed as juice. It is then sieved, mixed with the concentrated juice, and stored in dark-glass bottles. The bottles are stoppered with corks to stop air entering. Care is taken to make sure there is not too much air inside the bottles, and once they are opened and partially used they are decanted into smaller bottles.

Applying the compost preparations

A compost pile of around 40 m³ (52 yd³) is produced each year on Uppmälby. When preparing the heap, holes 60 cm (24 in) apart are made into the upper third of the heap using a wooden stake. Into them the individual compost preparations are inserted – one portion (1 tsp) in each hole. Twelve portions (dessert spoonfuls) of valerian preparation is stirred in about 15 l (4 ga) of rain or lake water. Holes are made in the compost from above and about 250 ml (8½ oz) is poured into each hole. The rest of the valerian preparation is then sprayed over the compost using a toilet brush. Once it has been prepared, a thin layer of soil is sprinkled over the heap before covering it with a blanket of old grass or hay. On top comes a piece of fleece in order to stop the hens scratching. The heap is usually prepared twice.

Burying and storing practice

The stuffed organs are buried in the topsoil. A wooden plank is laid over each of them to mark its exact position and to protect it from damage when it is being dug out. There is a place where each preparation is buried in the garden and each one is buried every year in the same place. The site is marked with a wooden stake and with the name of the preparation on it. In 2007, a new underground root cellar was built for storing vegetables, and the preparations are also stored there. It took a while for the climate inside to stabilise – it has now settled down to a more or less constant 5°C (41°F) and 95% humidity, which, according to Harald, is good not only for storing vegetables but also the preparations. The compost preparations are stored in clay pots inside a plastic crate lined with foil. Peat is beneath, and surrounding the clay pots, on top is a plastic sack filled with peat. All the preparations are evenly transformed, have no smell and are slightly moist (but not wet). Harald and Sonja store the preparations for a maximum of three years.

Preparation box at Uppmälby.

Derived preparations and other applications

Horsetail tea

A tea is made from dried horsetail by simmering it for an hour in water. This tea is used prophylactically against fungal attack. Horsetail tea is added during the last 15 mins of stirring the 501 prior to spraying it on potatoes. Harald and Sonja have had good experiences using horsetail tea on onions.

Stinging nettle extract

A bucket of fresh plant material is covered with cold water. In the first two days it has a sweet smell and can be used against aphids. Left for longer it can be used as a foliar feed, for which it should be diluted 1:10 with water.

Summary

Harald and Sonja are quiet, double revolutionaries. Quiet, because they follow their own path without making a song and dance about it. Revolutionary firstly, because the path they have chosen for themselves diverges significantly from what would generally be considered sensible, practical and economic in their region: they have been able, with modest means, to create a viable mixed farm. Revolutionary secondly, because with regard to preparation making, they have also found their own solutions. Steiner's indications remain important, but have been adapted to the needs and resources of their own farm. Their decision to use the organs and manure from their own sheep feels both true and consistent to them.

The important and clearly visible aspects of Uppmälby include the use of the farm's own resources, the conscious non-rationalisation and non-mechanisation of processes, and a conscious decision not to specialise. The diversity that has been maintained and cultivated on the farm because of these decisions, gives the farm a very special character. Harald and Sonja see in the interaction of these various elements, the uniqueness of biodynamic agriculture. In a similar way they experience the interaction between the various preparations as an effective whole.

It is very clear throughout the farm that the preparations play a very important role. The stirring barrel stands in the centre of the farm, right next to the sheep enclosure where both of them spend a lot of their time. The places where the preparations are buried are in the midst of the herb garden and therefore in the centre of things too. The path to the vegetable store is simultaneously the path to the preparation store. The preparation box, the animals, the structure of the farm as well as their own well-being, is looked after by Harald and Sonja in a loving, conscious and attentive way. Uppmälby is a highly individualised, yet simply constructed, mixed farm with a very unique approach to the preparations.

7. The Zeeland Group, Netherlands

Anke van Leewen, Dr Ambra Sedlmayr

Introduction

Biodynamic farms in the Netherlands have different options for obtaining the biodynamic preparations they need. If farmers don't make their own preparations they are able to buy them; previously volunteers made them for the Biodynamic Association, since 2013 the production and sales have been taken on by a farm (degroenenhof.nl). There are also a few preparation-making groups where farmers join together to produce the preparations. The groups are relatively small, with between four to eight farmers participating in each group.

The group based in the province of Zeeland, in the south-west of the Netherlands, was chosen for a case study on the recommendation of the Dutch Biodynamic Association. This group makes preparations for four farms in Zeeland. The managers of two of the farms – Boomgaard ter Linde and De Ring – participated in this study. The preparation work within this group is essentially supported by a core of experienced individuals who have been working in biodynamics for over twenty years. These are Helen Korstanje, Piet Korstanje and Margreeth Mak.

Dr Ambra Sedlmayr and Anke van Leewen visited the group on September 29 and 30, 2014. The farms Ter Linde, Boomgaard ter Linde and De Ring were visited. A group interview on preparation making was conducted and an in-depth interview with Margreeth Mak. The researchers also participated in the stirring and application of the 501 by the group on the De Ring farm. They also participated in the eurythmy exercises led by Boudewijn van Sluijs, and in practical exercises to perceive the effects of the 501 preparation using formative forces research methods, led by Frans Romeijn.

Farm portraits

The farms of all the preparation group members are located in Zeeland. Zeeland is a peninsula consisting of three former islands that have been joined since the Middle Ages. The farms visited are all located on these former islands. The soil in most places is very sandy and sometimes covered with a layer of about 40 cm (16 in) of clay.

The climate, strongly influenced by the North Sea, is temperate and maritime. The average temperature in winter does not fall much below 5°C (41°F) and in summer stays around 18°C (65°F). Days with a temperature above 27°C (80°F) are rare, as are days with frost in the winter. Average rainfall is 745 mm (30 in) for the city of Vlissingen, and is fairly equally distributed around the year

A short overview of the three farms managed by members of the preparation-making group follows.

Ter Linde Farm

Ter Linde has been a biodynamic farm since 1926. In the 1930s Dr Ehrenfried Pfeiffer, a close collaborator of Rudolf Steiner and an early developer of biodynamic agriculture, worked on this farm. The farm works together with another biodynamic farm, named Wilhelmina hoeve, to produce hay and raise cattle. Ter Linde farm has about 40 ha (100 ac) of grassland and 95 milking cows. Farm enterprises include a cheese making dairy, a shop and a small camp site. Most of the preparations made by the Zeeland group are produced on Ter Linde and they are stored there as well.

Boomgaard Ter Linde

Boomgaard Ter Linde is a fruit orchard, situated on the land of Ter Linde. It is managed as an independent business by Helen and Piet Korstanje and was founded in 1991. Prior to this Helen and Piet Korstanje were running another farm. Altogether they have some thirty years of experience with biodynamic fruit growing (boomgaardterlinde.nl).

Boomgaard Ter Linde has 17.5 ha (43 ac) of apple and pear trees. Several years ago, Piet and Helen participated in a formative forces research project and ran some scientific trials in their orchard. Their main aim was to discern consistent effects of using biodynamic methods. These trials consisted, amongst other things, in planting an area of fruit trees with varying distances between them, resulting in a more extensive plantation. The effect of this on fruit quality was studied.

Apple trees at Boomgaard Ter Linde.

De Ring

De Ring is another fruit orchard and has been managed, since 1992, by Margreeth Mak and Harald Oltheten in a village on one of the Zeeland peninsulas. The farm covers 4.5 ha (11 ac) and is planted with a wide range of soft fruit, including blackberries, raspberries, black and red currants, and gooseberries, as well as various kinds of plums and apples. The fruit is harvested and exported to different countries in north western Europe – mainly to a box scheme in the UK. Some fruit is also processed on the farm.

How the group is organised

The preparation-making group in Zeeland has been in existence for over thirty-five years. The group began by producing preparations for the Biodynamic Association of the Netherlands. Today it produces preparations for the four farms in Zeeland farmed by group members. Margreeth Mak has been the leader of the group for about two years and organises the meetings. She took over from Helen Korstanje. The group meets twice a year in spring and autumn to make the preparations. During these meetings the participants study the effects of the preparations together, in order to deepen their understanding of them. The group is open to anyone interested, such as farm workers and people who are generally interested in anthroposophy. For the last two years the preparation-making group has been working together with a group led by Frans Romeijn researching formative forces. Formative research exercises have been part of the programme of three meetings so far. The group aims to invite different contributors to their meetings, and changes the contributors after every three or four meetings. In the future, it is planned to work with the artist Paul van Dijk to experience the essence of water.

The responsibility for preparation making is shared between the farmers of the preparation group. Each of the farmers is in charge of one or more of the preparations. The person responsible has to provide the ingredients and animal sheaths for the group. For example, Helen Korstanje collects dandelion flowers and gets the mesentery. Margreeth produces the valerian preparation. Some of the decisions are made collectively, such as where to bury the preparations, but others are made by the individual concerned.

How the work developed

In the time before the BSE crisis the group was able to slaughter one of its own cows in the autumn and use the fresh organs for making preparations. From their account, it would seem they had a close and direct relationship to the preparation work at that time. The legal restrictions on using animal organs brought about by the BSE crisis in 2001 (ec.europa.eu), caused great insecurity among biodynamic farmers in the Netherlands concerning work with the preparations. With restrictions on the use of animal by-products, many farmers now had to buy preparations from Switzerland or elsewhere. This caused a great deal of dissatisfaction. Some of Piet Korstanje's colleagues even quit the Demeter certification scheme, finding the use of preparations from abroad incongruent with the biodynamic approach.

Another issue that caused an uproar in the Dutch preparation-making world occurred between 2003 and 2005, when the application of biodynamic preparations was made a requirement for certification by the Demeter Standards. It was argued that working with the preparations should remain an individual's free choice. Some Demeter certified farms at the time also had no regular preparation practice in place and had to make a decision.

The biodynamic community in the Netherlands therefore had to deal with these two changes – the restriction to the use of animal organs, and the new obligation to use the preparations – within a short time. Piet Korstanje remembers: 'I think a lot of things came together at that time, what with the BSE crisis and the problems we had with using the preparations and the obligation from the Demeterbund to use the preparations – things were chaotic. We thought about different organs for the preparations because we felt that, if you can't use the organs in a legal way, maybe it is also a sign that the time is right for finding new organs. There were a lot of people in our surroundings who had problems with the organs, because they were vegetarians'. There was a feeling at the time that substitutes for the animal organs needed to be found, so that work with the preparations could continue into the future. For Piet, the plants seemed to be the most important part of the preparations. So he asked himself: 'Isn't there another way to deal with the flowers, because the flowers are the main parts of the preparations? Can't you use the flowers in a different way, so that they become preparations as good as the old preparations? I had also this feeling, maybe we need new preparations and people are more interested in using new or renewed preparations, instead of using the old preparations that are illegal and which you have to use.' During this period, Helen Korstanje started to make experiments using silk and beeswax to replace the mesentery for making the dandelion preparation.

This was a period of insecurity and confusion that weakened work with the preparations across the country. There were also a lot of changes within the Zeeland preparation-making group at the time, caused by a change of staff on Ter Linde farm. Now, however, some years later, a new motivation to engage with the preparation work has been found within the more experienced core group that continued. Discovering formative forces research methods gave them a tool for developing their individual experiences and increasing their understanding of the preparations. This was key to their further engagement with the work. Working with eurythmy as a means to understand natural processes has engendered additional enthusiasm. The aim of the group is now to deepen its understanding of the preparations. This understanding they believe is essential nowadays in order to motivate people 'to do this important work.' Former participants who did not want to engage in the formative forces research have left the group, while others have been attracted because of it. Some new biodynamic farmers joined the group in the autumn of 2015.

For the group, the social aspect of striving and working together is very important. Members feel how the warm openness of the group helps them to trust in their own experiences and observations. Group members do not always have the same opinion but can accept different positions. During the meetings, and after every experiential exercise, each person is asked to share their thoughts and experiences. Frans Romeijn says of the group work: 'We are a research community. Out of anthroposophical understanding it is important that we work together. It enables spiritual beings to connect with us.'

Margreeth Mak said that she would like to involve more people in the preparation group, even though it might change the dynamics of the group. Not everyone in the group, however, feels the same. Margreeth believes that the intention of wanting to work with and understand the preparations is more important than carrying it out in a certain way.

Social interaction and formative forces research are, therefore, providing the foundations upon which work with the preparations can develop within the Zeeland preparation group today.

Zeeland preparation-making group during formative forces research.

How Margreeth Mak found her way into preparation work

Margreeth Mak was chosen as an example of the journey taken by an individual of the preparation group with regard to the preparations. She was chosen because Margreeth is the current group coordinator and was willing to share her personal experiences with the researchers.

Margreeth was born in 1966. She grew up in the middle of the Netherlands, in Veenendaal, as the youngest child in a family with four children. For her family, church was very important. In her childhood,

Margreeth spent a lot of time on her own out in nature, spending hours in contemplation of plants and water. When she was nine years old, she heard about a so called 'Rudolf Steiner School' for the first time, and without even knowing anything about it, knew that she wanted to go to this school. Because she was aware that her parents would not take her out of the ordinary school, she did not talk about this wish to anybody. When she was twelve, Margreeth came across anthroposophy again and was very interested. By chance she visited the Warmonderhof biodynamic agriculture college, and immediately knew that she wanted to study there. She explained: 'I learned in my youth that there is a world above us and we are not in it, we only know about it. We believe it must be there, but during our normal life we cannot do anything with it. In biodynamic agriculture it is possible to directly work with it, this is what I saw at the Warmonderhof.' It was during this visit to the Warmonderhof that Margreeth heard about the biodynamic preparations for the first time.

Margreeth kept her early decision to study at the Warmonderhof herself until she finished school. Then she felt that she had to develop a connection to the land, 'to learn to love the earth' as she said, and went on to do an internship on a biodynamic farm in France. Later, she made her wish come true and studied biodynamic agriculture at the Warmonderhof. After her studies, Margreeth and her husband started the fruit orchard De Ring.

Margreeth had learned about the preparations during her training at the Warmonderhof. She did very little on a practical level for the first ten years of the farm's development, however, because she was too busy raising her children. When she later was able to put more time into the fruit orchard, she took on the preparation work as well. She felt somewhat insecure at that point about what she was doing and so she joined the preparation group that was then being coordinated by Helen Korstanje. This helped her to get back into preparation work.

Margreeth then had an experience while stirring the 500 on her own that was of great significance to her and which motivated her to deepen her work with the preparations. She felt the presence of a being for the first time while working on her own. She felt very connected to this being and even had the feeling that she had given the impulse for its formation. At the same time, the being seemed to be an independent individual for her. During this encounter she had a kind of timeless feeling. She felt in harmony with nature.

Helen was looking for someone who could take on the organisational aspects of the preparation group and, after a while, Margreeth agreed to take it on, even though she did not feel very confident in her work with the preparations. In bringing together the preparation group with the group working on formative forces research, Margreeth felt that she had found a way to build a good foundation for her own and the group's preparation work.

The Zeeland group's understanding of the preparations

Most of the group members feel it is very important to make their own preparations. They feel strongly about having preparations produced in the environment of the farm. Helen Korstanje said that the connection that farmers have, and the intention they give to their land through their work with the preparations, is important for the beings that help plants and animals to grow.

Margreeth has often thought about the influence a person has on the preparations being handled. She wondered: 'What effect of the preparations is due to human influence and what comes purely from the preparations? What is the role of a person, of me or Frans or someone else, what do we add to the water? Am I allowed to add something to the water, to be involved, or do I have to stand back? What is the influence of my thoughts and my feelings? Is there an influence on the preparations when I doubt their effects? Should I hold my doubts back? Are there thoughts that I should add? What is my role?' Margreeth believes that the thoughts and feelings of the person handling the preparations do have an impact on them. She therefore feels that personal development work is essential in order to allow ever better and purer intentions to flow into the preparation work. She said: 'When I am stirring alone, I am very attentive to all my feelings and thoughts ... Do we bring all this in? Yes, I am sure about it. So it is my big mission to become a better person, and it is especially when I am stirring that I am aware of this.' Margreeth believes that it is always herself that she has to face in the work on the farm. Personal development work is therefore important for her.

The group's understanding of the animal sheaths was furthered by an experiment made by Helen Korstanje during the time of the BSE crisis. Due to the BSE crisis, access to animal organs was restricted and Helen Korstanje started to question the use of animal organs. Before the BSE crisis, Piet and Helen Korstanje had always had access to fresh organs from biodynamic cows to make the preparations. With the changing regulations and considering the growing vegetarian movement, Helen thought that perhaps the time of making preparations with animal organs was over. She asked herself whether humans really have the right to take the life of an animal and to use animal organs to make the preparations. Helen tried to find alternatives and made a trial with the dandelion preparation. She made a skin of silk and beeswax sewn together to form a pocket for the dandelions as a substitute for the mesentery. One side was left partially open so that the dandelion could breathe and was not too enclosed. She buried it like the other preparations and when she took it out, it looked like a normal dandelion preparation. It did not keep in storage however and turned to dust half a year later. Helen tried this for two years with the same result. She was very impressed by the evident role of the animal sheath. This experiment made her conclude that her new preparation did not work and that the 'classic' preparations 'really must be something.'

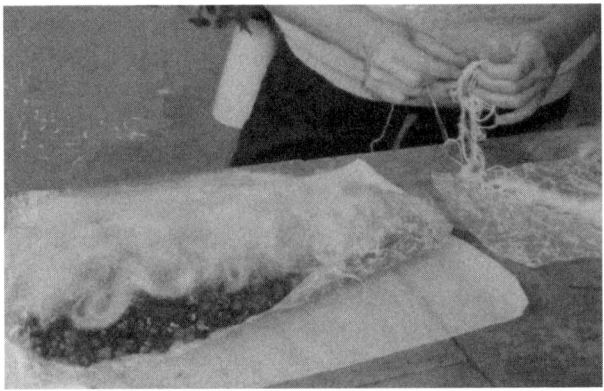

Dandelion trial.

Regarding the effects of the preparations, these are mainly investigated with the help of the formative forces research group. Piet Korstanje had worked with biodynamics for many years before having his first, deeper experience with the preparations. Having applied the 500, he once had the feeling that something was coming out of the ground into the plants. With the 501 he noticed another type of movement, one that connected the leaves with the sky.

The group discusses the quality of the preparations when they are dug out. They assess their appearance and their smell and compare the findings with the impressions of previous years. Some group members take notes every year. The preparations in the store were relatively dry, although they were not purposefully dried. The plants were not totally transformed and the original flowers could still be recognised.

Personal experiences with eurythmy and formative forces research
Anke van Leewen

During data collection for the present case study, the researchers participated in a group process similar to those taking place at regular group meetings. The first point of the programme was eurythmy with the Steiner school teacher Boudewijn van Sluijs. The group tried to step into a deeper understanding of life processes through making them visible and experiencing them with the help of eurythmy. In this way, plant germination was explored. The group observed and discussed the process step by step. Which gestures are involved and which directions? Where are the turning points?

The attentive observation continued on another level with the use of formative forces research methods. In both disciplines, eurythmy and formative forces research, the body is involved and used as an instrument. While in eurythmy it plays an active, visible role, in formative forces research it is mainly perceptive and its use more subtle. The formative forces exercise took place in the fruit orchard of De Ring. Each group member walked through two lines of blackberries and focused on the changes in their own bodily perception. When they shared their results, it was found that most of the group members felt a density between the plants and an upward-directed movement. There were also perceptions on other levels, like the feeling of warmth, seeing light, or feeling sad because the plants were cut.

The next part of the meeting consisted of the group observing and evaluating the effects of stirring and spraying the preparations in the orchard. The shared activities encouraged everybody to participate in a vivid exchange.

Frans Romeijn

Frans Romeijn aims to demonstrate how the etheric forces can be experienced as a reality. He is using the approach of Dorian Schmidt described in his book *Life Forces – Formative Forces*, whereby one's own etheric body is used to perceive the kind of forces at work in a plant, food or water. The three senses that are involved in the perception are named by Steiner as the sense of life, the sense of movement and the sense of balance.

When Frans is working with biodynamic farmers, he focuses on the effects of preparations. From his point of view a lot of farmers have difficulties in experiencing the effectiveness of the preparations when working with them. He shows them how to do practical exercises that can lead to them having their own perceptions. He believes that it is important for farmers to experience the effects of preparations in a direct way. Frans hopes that he can use this method to help motivate farmers to work with the preparations.

Preparation practice
Field spray preparations
Horn manure preparation (500)
The manure for making the 500 comes from the Ter Linde farm. It is collected on the pasture and in the cow shed. About 200 horns are filled on a date around Michaelmas, together with people who are invited to help and join the group. The place for burying the horns is always near the cow shed close to where the cows pass by. It is a central place in the farm. The depth of burying is about 40 cm (16 in). One year, the 500 disappeared over winter. When it was dug out the horns were empty.

Different group members use different quantities of the 500 for stirring. Margreeth Mak uses about 100–150 l (26–39 ga) of water and the content of three horns for 4 ha (10 ac) of land. Piet Korstanje uses the content of about two horns per hectare. On both of their farms the 500 is applied twice a year. When the soil is cold and wet in the springtime, for example, Margreeth also adds about 15 ml (½ fl oz) valerian preparation 15 mins before she finishes stirring the 500. Her aim is to give warmth to the soil.

Horn silica preparation (501)
Margreeth Mak is responsible for producing the silica preparation. She buys fine silica powder from Weleda. In order to fill the horns, she mixes the silica powder with rain water to get a paste with a fluid consistency. The filled horn is buried traditionally next to the bee house and taken out around Michaelmas. The finished preparation is stored in a glass jar in front of a window. Margreeth uses very small quantities of the 501 for spraying the fields, only about 1 g (one knife point) per hectare.

In previous years the group stirred the silica paste for 15–20 mins before filling it into the horn. Margreeth stopped this practice, because she did not have a good feeling about it and because the stirring before spraying seems more important to her. But other people in the group feel differently. Helen Korstanje for example argued: 'When you stir silica for maybe 20 mins, you make a connection to it, and this connection makes the silica even more silica-like.'

In the experience of the group, the 501 can be kept for many years.

Stirring the spray preparations
Certainly each group member has an individual approach to stirring. Margreeth Mak has a special way of stirring. She stirs rapidly during the

final stage of building up the vortex, then she leaves the stick in the water (hanging from a tree) until the movement of the water has slowed down somewhat, then she starts stirring in the opposite direction. The movement of the water when changing direction is gentle in comparison to how others do it. Margreeth explained that she 'hates to make a chaos' in other areas of her life and that this influences her approach to stirring.

During the visit, an observational study was made of the stirring and spraying of the 501 using formative forces research methods and accompanied by inner observations. Stirring took one hour, everybody stirred for some minutes. It was the first time the group had stirred together.

After stirring the 501, the group shared their experiences. The group members had observed a warm dome of light around the stirring barrel. The group discussed and agreed that the boundary of this phenomenon extended to some 7 m (23 ft) around the barrel. They had the impression that after being sprayed, trees were 'happy' and spread out their branches and leaves. The sprayed preparation was also felt to radiate out into the surroundings. In this way other trees could benefit from the preparation. The effect of the 501 was observed not only as being positive, but also as causing a certain tension and tightness.

Later, the group discussed their experiences of stirring together. Some people experienced the sharing of stirring as positive. Margreeth felt supported in this normally lonely task that sometimes arouses the suspicion of neighbours. Others commented that the extra chaos created by different people stirring must be positive, since in this way 'more new influences can come in.' Others considered that the diversity of experiences and intentions that people brought to the stirring was positive. Piet was more cautious, as he felt the experience of stirring on one's own is more concentrated and focused. Another person noted that when stirring alone, one can feel how the water changes throughout the process and that a new 'force' is perceptible after an hour of stirring.

Compost preparations

Yarrow preparation (502)

Yarrow is collected by Margreeth Mak on her farm and nearby. The bladder is bought from the Biodynamic Association. She uses the fresh flowers and fills the bladder in July. The bladder is hung up next to the bee house, but with a protection against birds to prevent damage. This bladder is then buried around Michaelmas.

Chamomile preparation (503)

The work to collect chamomile flowers is shared between several group members. The flowers are dried on a nylon net and stored in paper bags. The intestine is bought from the Biodynamic Association. The flowers are moistened with chamomile tea before being placed into the intestine. The intestine itself is moistened with lukewarm water before use to make it supple again. In 2013, the intestine was buried in Margreeth's fruit orchard, but roots were found growing strongly into the preparation and so the group is now looking for a new site.

Nettle preparation (504)

Nettle is buried every year in June and taken out a year later at about the same time. Whole plants are harvested, right at the beginning of flowering. After mowing, the plants are dried for three or four days and then buried with a bit of peat around them. The nettle is buried at a quiet place, situated in the middle of the pear trees in Boomgaard Ter Linde and within a place reserved for birds, insects and other animals.

Oak bark preparation (505)

The oak bark is collected on the Ter Linde farm. It is cut off the tree with the help of a planer and ground afterwards. For producing the oak bark preparation a bovine skull is used. Currently, the skulls come from an organic farmer, who keeps a rare old Dutch breed. The skulls are often damaged, as animals are killed with a nail gun on their head. Since skulls are difficult to obtain, the same skull is used for four to five years. A blue plastic barrel is buried in a place where rainwater can be funneled through. The skull is then placed inside this barrel in such a way that it emerges 5 cm (2 in) above the surface of the soil and water contained in the barrel. This is to prevent the preparation getting a bad, putrid smell. The barrel is finally filled with plant material and rain water. Piet Korstanje is responsible for producing the oak bark preparation.

Dandelion preparation (506)

Collecting dandelion flowers is a task shared between several people. The flowers are dried for use in autumn. Helen buys the mesentery from a butcher. For assembling, it is placed in lukewarm water to become supple. The flowers are moistened with dandelion flower tea. The moist dandelion flowers are placed on pieces of mesentery that are then formed into a ball and sewn together. Helen Korstanje buries the dandelion pockets at about 40 cm (16 in) deep into the soil next to the bee house.

Although the mesentery is the organ in use, the group is not sure if it would be better to use the peritoneum instead.

Valerian preparation (507)

Margreeth is responsible for the valerian preparation. She collects valerian flowers on the polder (the new land that has been created by putting up dykes in the sea), and on her farm, since she has planted some plants there. In the future, she hopes to be able to produce all the valerian flowers she needs for making the preparation herself. For making the valerian preparation, only the petals are used. Margreeth produces a pure juice with a juice press. The flowers are wrapped in a cotton cloth and set under pressure. One litre (one quart) of extract is needed per year. She stores it at home in a brown glass bottle and it stays fresh for a long time.

Applying the compost preparations

In Margreeth's orchard, compost is treated twice with the preparations. The preparations are arranged in a line along the compost heap, with 3 m (10 ft) between each preparation. She uses a marble-sized piece of each preparation. The nettle preparation is always placed in the middle between the other preparations. She stirs a drop of valerian juice in lukewarm water for 15 mins before it is applied on the compost heap with a knapsack sprayer. The concentration she uses is 5 ml (1 tsp) of valerian preparation in 7 l (2 ga) of water.

On Ter Linde farm it is not only the compost heap that is inoculated with preparations, but the manure receives its first treatment while still in the cow shed.

Burying and storing practice

The stuffed organs for preparation making are buried in autumn around Michaelmas and dug out in late April or beginning of May. The group used to have specific places to bury the preparations. These were on more natural, undisturbed parts of the farms. But the group encountered problems when tree roots were discovered growing into the buried material. Currently, the group is looking for new places, in open fields, to bury the preparations. The criteria for choosing a spot to bury the stuffed organs are different for each preparation; the aim is to find places that support the special quality of each preparation.

Thought is given as well to how the maturing preparations should be distributed on the farm organism. One idea was to bury all preparations next to the farm centre, so that they are in the heart of the farm. But, finally, the choice was made to bury them all further away from the farm centre. On the one hand the group want to separate the preparations from each other so that they can develop their independent qualities, undisturbed by the other preparations. On the other hand, group members think it would be good to give more attention to some places further away from the farm buildings. In this way the heart of the farm would be expanded.

Storing the preparations

Horn manure and all the compost preparations, except for valerian, are stored on the Ter Linde farm. There is an old root cellar with a balanced climate that is now solely dedicated to preparation storage. The preparations are stored in glazed ceramic pots, which are then placed in peat that fills the space between the cellar wall and a wooden board. The ceramic pots stand in peat but are not fully surrounded by it.

Summary

For most members of the preparation making group of Zeeland, a personal relationship to the preparations is very important. This includes being actively involved in preparation making. There is a great wish to gain a deeper understanding of the preparations, because they believe that only by understanding them themselves will they find the strength and motivation to continue the work.

There was a lot of change in the group composition previously, but a core group of people has ensured the continuation of the work. Four years ago they found new motivation for their work through the formative forces research group of Frans Romejn. Formative forces research is currently the focus for their studies. The group has a lively culture of discussion and mutual tolerance.

Their relationship to the preparations was challenged by the difficulties arising from the BSE crisis. At that time they tried to find substitutes for the 'old preparations' that would allow them to work without animal organs. These experiments demonstrated to them the importance of the animal sheaths.

Some of the group members feel there is a connection between elemental beings and preparation work. Margreeth Mak described an encounter with a being while she was stirring and of her feeling of responsibility for this being. Margreeth recognised a need to develop herself, because she believes that there is a significant influence coming from the person doing the work on both the preparations and the farm.

8. Carlo Noro, Labico Farm, Italy

Dr Maja Kolar, Anke van Leewen

Introduction

Carlo Noro is a professional preparation maker based in Labico, 45 km (28 miles) south of Rome, Italy. On his farm he produces biodynamic preparations that are sold throughout the country and are also exported to other European countries. Together with his co-worker, Michele Lorenzetti, Carlo is currently conducting research into the microbiological activity of the preparations.

Labico has cold and rainy winters and warm, dry summers. The climate is classified as Mediterranean (humid winters and dry, hot summers), with a warm, temperate, moist, forest bio-zone (chinci.com).

Dr Maja Kolar and Anke van Leewen visited Carlo Noro on October 9 and 10, 2014 on the recommendation of the Italian Biodynamic Association and Demeter, Italy. The interview was translated consecutively from Italian to English by Michele Lorenzetti, a biodynamic consultant and a co-worker of Carlo. Two interviews were conducted and two farm visits were made in order to collect the data. A visit was made to the farm where Carlo lives and works and another to a large farm to the North of Rome, where Carlo collects manure for the 500. No preparation work was done during the visit.

Farm portrait

Carlo took on his farm in 1998 in Valle Fredda, which means 'Cold Valley'. There is usually a long period of intense icy weather in winter, and this was one reason why Carlo chose this place as he considers such conditions to be beneficial to preparation making. Before Carlo bought it, the land had been an intensively managed vineyard.

The 5 ha, (12 ac) farm is on two sites. The site where Carlo and his family live has some arable land and greenhouses. The other site has a vineyard and olive grove and is located in Piglio, a small town on the slopes of the Scalambra mountain in the vicinity of Rome.

The average annual temperature in Labico is 14°C (57°F). August is the warmest month, with an average midday temperature of 29°C (85.5°F). January is the coldest month, with night-time temperatures averaging 1°C (35°F). Annual precipitation is 900 mm (35 in). Labico has a humid climate with precipitation fairly evenly distributed throughout the year (climate-data.org).

Carlo Noro's farm and vegetable fields.

The soil in Labico is a cambisol of volcanic origin. This is a moderately developed soil whose lower horizon (B horizon) has a different colour, or structure, to that of the mother rock (Carta Suoli Italia, 2012). It tends to be very acidic, with a pH value of 4. The soil on the farm has a fine and stable crumb structure with a high humus content and light brown colour. It smells like a mature forest soil and has a pH value of 6.

The main products of the farm are vegetables, herbs, wine, olive oil and fruits. The greenhouses cover an area of 300 ha (741 ac). A further 2 ha (5 ac) are dedicated to growing vegetables such as tomatoes, carrots, salads, peppers, potatoes, asparagus, courgettes, pumpkins and aubergines. There is no particular focus on crop rotation and companion planting, since, in Carlo's experience, a well-balanced soil always produces healthy and resistant plants. There are 200 olive trees and half a hectare (1 ac) vineyard growing red and white grape varieties. Most of these products are processed on the farm and are mainly sold to local customers via a twice weekly stall set up on the farm and several restaurants in Rome.

The biggest focus on the farm, however, is the making of biodynamic preparations. The preparations are sold to over 400 different farms in Italy and to some farms abroad. Biodynamic courses and projects concerned with environmental sustainability, arranged with civil society and other stakeholders, provide additional sources of income.

Carlo used to have sheep, but because of the limited space available on the farm he decided to sell the sheep and dedicate the available space to vegetable production. The only livestock he has at present are 22 organically managed beehives, owned by a local beekeeper.

The place is managed as a family farm. Carlo is the owner and two of his sons are working on the farm.

First steps in preparation practice

Carlo was born in 1951 in Piglio, Italy. When he was twenty he met an anthroposophical doctor who encouraged him to find out what he would really like to do in life. Carlo realised that he would like to become a farmer, but not in a conventional way like his father, who used a lot of chemical inputs. Carlo wanted to find another approach to agriculture. The anthroposophical doctor, who had become a friend and teacher of Carlo, suggested that he should find out about biodynamic farming. He started to study anthroposophy in 1977 while he was working in a bank. Carlo felt that he was gaining a lot from anthroposophy for his own development as a human being, and he began to think about how to give back something of what he had received from life and anthroposophy.

The more he learned about biodynamics, the more he realised that this was the way of life he wanted to follow. When he read the fourth and fifth lectures of the Agriculture Course for the first time in 1981, he knew immediately that his mission would be to work with the preparations, since they resonated so strongly with his understanding of life and agriculture.

How the work developed

After reading the fourth and fifth lectures of the Agriculture Course in 1981, Carlo started to make the preparations by himself without any advice. In the same year, a Section for Biodynamic Agriculture was founded in Rome. Many people were interested in biodynamics but the group soon disintegrated because, according to Carlo, nobody was able to transform the theory into practice.

Nevertheless, Carlo's impulse stayed strong. His main question at that time was why everybody was talking about drying the preparations. He asked himself what happens to their living quality when the preparations are dried. He felt that their vitality would dissipate through drying and that the preparations would no longer be active. Carlo wondered why nobody seemed to agree with his idea of not drying the preparations. He thought he was mistaken and that maybe he had not understood something correctly. This was a difficult time, living with this dilemma. His way of making the preparations was to use fresh blossoms and not to dry the preparations at any stage. He asked himself: 'If the end result of the process is moist, why should it then (at any stage) be dried?' This was the main question he was not able to find an answer to and he almost gave up.

In 1982, Carlo met Alex Podolinsky when he visited Agrilatina (in Lazio in Italy) for the first time. Carlo recounted how Alex had prolonged his stay in Italy in order to visit Carlo's place with his assistant, because he was so impressed with the preparations that Carlo had shown him. Alex took a different approach to the making of the preparations than the people Carlo had met hitherto. It was much closer to his own ideas and Carlo felt he could learn a lot from Alex.

An important moment came for Carlo when he realised that he did not have to follow the Maria Thun Calendar. One day in June, Carlo was in the vegetable garden looking carefully at the flowering chamomile but not harvesting it. His father came and asked him why he was not harvesting the chamomile. Carlo answered: 'Because it is not the right time, it is not written in the calendar.' But his father said: 'They are flowering, they are in blossom, the conditions are perfect, what more can you expect?' Carlo realised that he had been taking the Maria Thun Calendar as if it were a 'holy book'. He saw wisdom in his father's suggestion and decided to follow it. It dawned on Carlo that many people who preferred to dry the preparations did so because they were using older flowers that had already started forming seeds, and that these would germinate if the preparations were stored in a moist state. The question for Carlo became how to make the preparations in such a way that they would not degenerate?

Following their meeting in 1982, Carlo started to work more closely with Alex Podolinsky and, as a result, many of his doubts and uncertainties were clarified. Alex encouraged Carlo to keep the preparations moist, an approach that made sense to him. From then on Carlo always kept the preparations in a moist state like Alex did in Australia. Carlo feels that biodynamics will only work if the preparations are kept moist.

Alex asked Carlo to make preparations professionally, but Carlo was not yet able to leave his job at the bank because it provided an income for his family. He told Alex, however, that in four or five years time he would leave his job and find a new way of making a living with biodynamic farming.

In 1998, this became possible when Carlo found a suitable place for making preparations in Labico, Valle Fredda. The soil there is acidic and he could cultivate all the plants needed for the preparations. From 1999 to 2000 onwards, Carlo began making the preparations on a big scale for sale.

Carlo Noro's understanding of the preparations

The effectiveness of the preparations

It was very important for Carlo to be able to see the effects of using the biodynamic preparations. Even though he felt it was his destiny to work with the preparations, he promised himself, 'I will make the preparations but if I find they do not work I will not waste more time with them.' Carlo gradually noticed several effects of the preparations. Even though there appeared to be no immediate results, he found that after some time concrete effects were discernible. Of key importance for Carlo was the improvement in soil structure. He said: 'Another effect of the preparations has been the improvement of soil structure. Today, even heavy rains do not damage the soil. We can always find a solution if we work with nature.'

In Carlos' eyes, creating good soil is the main task of the biodynamic farmer and it provides everything needed for healthy plant growth. In his words: 'If any external input is needed on the farm, it is because something is missing or out of balance.' Carlo said that he has managed to bring his soil from a pH value of 4 to a pH value of 6 simply by using the preparations.

Quality preparations require professional engagement

Carlo believes that high-quality preparations are essential for good field results, and that dedicated people, with expertise and sufficient time to invest throughout the year, are needed to make the preparations and ensure they are of high quality. In Carlo's opinion: 'We should use the gift that Steiner gave to us. The work has to be very well done and it takes you a good part of the year.' Carlo is worried about the quality of the preparations being made at present and believes that top-down, concerted action needs to be taken at the level of the international biodynamic movement to improve their quality.

To produce high-quality preparations, great care needs to be taken at every stage and with all aspects of the work with them, because the preparations are alive. He explained: 'The life is more important than the theory. We have to be more conscious about how to make them. I know that life is inside.' Carlo also keeps in mind the person who will eventually be using the preparations and that they need to be effective where they are applied.

Carlo says that preparation quality can always be improved and that he is continually working to produce better preparations. Carlo finds that in working with the preparations, new ways of improving their quality continually appear. He said: 'Making improvements is a very dynamic process, with innovations and new insights coming unexpectedly.' The process of working with the preparations becomes, in this way, an aim in itself. To begin with, Carlo was very unsure about how the preparations work, but over time his accumulated experiences have convinced him of the effectiveness of the preparations he is producing.

Carlo has, together with Alex Podolinsky, found a way of judging the quality of the preparations. In their view, a complete transformation of the original materials is the hallmark of a high quality preparation.

Making sense of the biodynamic preparations

Alex Podolinsky showed Carlo the way he makes the preparations. Carlo and Michele have taken up this practice and developed it further in their own way. For Carlo, understanding the preparations requires an inner understanding of life and, instead of discussing theory, there needs to be a practical, conscious experience of working with the preparations. He said: 'One cannot understand them if everything is built just on theory. I can teach that inside 500 there are etheric forces. If I do not have this concept from practice inside myself, I cannot teach it.'

With regards to plant life, Carlo explains that, rather than developing quick-fix solutions based on superficial observations like conventional agronomic science does, by observing the plant and getting an inner understanding of it, one can find holistic solutions that are already immanent in the plant itself. He says: 'Be a friend with nature, it always gives us the solution.'

The social setting

The priority of Valle Fredda farm is always the production of preparations, even when it is time to harvest, sow or sell vegetables. The whole family is involved in the preparation work. Carlo's eldest son is helping him and he will take over the preparation work step by step in the future.

The community life of the farm is also organised around the preparations. When there is work connected with the preparations that needs to be done, like picking flowers, everybody drops their normal work and comes to help on the field.

Carlo does not speak a lot about the preparations to other farmers, which is something he regrets. He does, however, run courses open to everyone, which are attended mainly by people who want to start biodynamic farms or farmers who are wishing to convert to biodynamics. Carlo prefers people coming to his farm rather than traveling to them. Farmers, researchers and people from the Biodynamic Association are in this way meeting his farm when they come to receive advice.

Preparation practice

For Carlo, 'the preparations are at the heart of biodynamic farming and biodynamic farmers should give most of their attention to them.' Carlo uses all of the preparations indicated by Steiner on his farm and, in addition, also the 500P. He used horsetail tea when his soil was not yet in 'full balance' he explained, but no longer feels the need to apply it.

Field spray preparations
Horn manure preparation (500)

Carlo collects the manure from another Demeter-certified farm, which is located one hour's drive away from his farm. He takes the manure directly from the pasture and only from adult, female animals. The pasture must have been sprayed with the 500 and the 501 before the cows come to graze. When the grass first starts to grow after the dryness of summer, and the animals start to eat the young grass, they produce a near perfect manure for making the 500. The best time for making the preparation is dependent on the weather of that year. The best time is normally between October and November, when the pasture is ready but neither too young nor too old.

Carlo Noro on the pasture with the cows.

Carlo always collects the manure in the morning. It is important to him that it is fresh and that microbiological activity starts inside the horn rather than on the pasture. Because of this and due to the large number of horns he has to fill each year (around 40,000), he spends several days on this other farm to collect fresh manure each morning. Carlo only wants to use manure that is produced from fresh grass and not from cows fed on hay. This is because Carlo believes that a cow's digestion works best with fresh grass and that hay is only a supplementary fodder.

Carlo has experienced some problems with the transformation process of the manure when using new horns. He thinks this is because the microbes needed for the right transformation of the manure are not yet present in the new horn. He has found therefore that horns used for the second time or more bring about the best results. He sees the microbiological processes taking place inside the horn as being essential for the transformation of the manure.

He buries the filled horns 20–120 cm (8–48 in) deep in the soil. In his opinion there is still a lot of microbiological activity at this depth. Carlo prefers to bury the horns in a place where there are frosts and where the cold persists for a while in the winter so that the horns can be sufficiently exposed to the forces of crystallisation. A good soil structure allows the crystallisation process to go quite deep into the soil. The level of humidity also affects the transformation process. If the horns are put into dry soil in autumn, Carlo waters the place to allow the process to start properly. If it rains too much the place needs some kind of cover, otherwise the transformation process will come to a halt.

When preparing the 500 for spraying, Carlo first of all warms the water to 30–32°C (86–90°F), not more. In his opinion 37°C (99°F) is too warm, as this temperature already inhibits some of the microbiological activity. To warm up the water, Carlo uses gas or wood fire. He does not like to use electricity when he works with the preparations. Carlo also warms up the 500 slightly. When both the water and the 500 are pretty much at the same temperature, he brings them together. A 20°C (68°F) difference in temperature between the two would be a shock for the microbes in the 500. This practice is called acclimatisation and is a detail piece of advice from Michele Lorenzetti, a biodynamic advisor, biologist and winemaker who has been working collaboratively with Carlo for fifteen years and doing scientific research in biodynamics.

Carlo suggests that large quantities of 500 need to be applied during the conversion period: 1 kg (2 lbs) per hectare each year distributed over four or five applications. But he also thinks that it is a mistake to prescribe precisely the same quantity for each place, since soil and other conditions vary so much from farm to farm.

There is a period in spring and in autumn when the microbiological activity of the soil is higher and this is when Carlo applies the 500, choosing a time after rain when the soil is completely wet. His practice is to spray the 500 by hand, even on the big fields. This is because entering the field with machines when the soil is wet could cause damage.

Horn silica preparation (501)

Carlo travels to the Apennine mountains in order to collect around 20 kg (44 lbs) of quartz each year. He selects crystals that are transparent.

To grind the crystals, Carlo uses a machine specially constructed by his friend Gianni Montanari. It contains a ceramic cylinder, a turner and a grinding element that is almost as hard as a diamond and which crushes the quartz in a vertical movement. The machine crushes the quartz crystals for several hours and it is then taken out of the machine and mixed with water to make a creamy paste. This paste is then put back into the machine and is ground for another six to eight hours, depending on how large the quartz crystals were at the beginning of the process. At the end, a milky quartz emulsion is obtained.

The horns are filled with the milky liquid in spring and placed 20 cm (8 in) deep into the soil. No consideration is given to the Maria Thun Calendar. What is most important for Carlo is that the soil is not dry but rather a little moist. Carlo puts the horns into the soil with their opening facing upwards. He believes that the horns containing quartz should be exposed to both soil and sun. In Carlo's opinion, his task is to give the preparations the possibility of evolving in the right way. The preparations are taken out of the soil in late autumn.

Carlo Noro explaining how the 501 is made.

Carlo believes that the 501 can be dissolved in water. His 501 is so fine, that one cannot feel the particles on the tongue. Carlo uses the 501 two to four times a year at the beginning of a plant's growth. The application is carried out using a quantity of 2–3 g (1 tsp) per hectare, between 6.00 am and 10.00 am, when the sun is already shining. This is because Carlo believes that it is important for the light to influence the preparation. Carlo described the process as follows: 'When it is sprayed, horn silica acts like a prism and separates out the colours giving monochromatic light to the leaves. This is why there should be light when horn silica is sprayed.' He later described how there is a similarity between this light process and the education of children: 'Monochromatic light is very important for the child in their development. The developing leaves of plants can take up all these single colours. A plant growing with too much nitrate receives unnatural colours, because the nitrate is toxic for the plant. If the plant is healthy, it is able to retain the colours of the whole surrounding cosmos.' Carlo also says that 'according to Steiner the colours form the character of a child. When the 501 is given to a plant, colours are also increased in the plant.' That is why Carlo thinks that the 501 should be applied at an early stage of plant growth when photosynthesis is still very active.

Stirring and applying the spray preparations

Carlo's advice is that stirring should be done by machines, as in his opinion that is the only way to ensure that the preparation is always of the same high quality. If the stirring is done by hand, its result cannot be known in advance and the resulting dynamised preparation can be either very good or very bad, depending on who stirred it. Carlo thinks that a person can only stir about 40–50 l (10–13 ga) by hand in the right way. Spraying a large farm would therefore require several people to stir about 50 l (13 ga) each.

Carlo uses an old spraying machine to distribute the 501. It produces a fine mist 6–7 m (20–23 ft) high.

Compost preparations

Carlo feels that the compost preparations play an important role in completing the processes begun by the 500 and the 501 in soil dynamics and plant growth.

In spring, before taking out the compost preparations, Carlo first takes a sample to see whether the microbiological processes have taken place correctly: for example, if the transformation of the preparation has happened. According to Carlo, this depends largely on soil moisture levels.

Yarrow preparation (502)

Carlo harvests wild growing yarrow from his farm. He collects the flowers himself, without help from others, because he can collect yarrow each day for several weeks. Carlo does not use the whole inflorescence but separates off the small flower heads and uses these for making the yarrow preparation. The flower heads are dried in the shade in a wooden hut and later stored in a glass jar.

Carlo uses dried stag bladders imported from Slovenia, as it is not possible for him to get fresh ones. He rehydrates the bladders with fresh water before using them. He then collects the water from the bladder and uses it to moisten the yarrow flowers so that all the substances and processes active in the bladder are brought to the yarrow. Carlo explained that the microbial activity that was previously active in the bladder now begins to work on the yarrow in the right way. The bladders are hung up under the corner of the roof over summer and buried in the soil in autumn.

Carlo pointed out how, after the process has been completed, the individual flowers can still be recognised. This is because the calyx is very resistant to decomposition. However, the flowers inside the calyx are totally transformed. Carlo believes that the calyx is so resistant because of the high potassium and calcium carbonate levels it contains.

Chamomile preparation (503)

Before harvesting chamomile on his farm, Carlo goes to the field on his own and picks all the flowers that are too mature (these are used for tea). Then he waits for the new flowers to emerge. Depending on the weather, it can take up to four days before the new flowers are ready to be picked. When they are ready, four or five people join Carlo and help with the picking, which might take the whole day. If left on the plant for too long, the chamomile flowers are no longer used for preparation making. Carlo developed this method, because when he first made the chamomile preparation, he found that the preparation did not last. He realised that when the flowers at the top of a flower head start blooming, those at the bottom are already setting seeds. The best moment to harvest is when half the flower head is not yet pollinated. This helps to avoid the problem of seeds germinating in the finished preparation.

In spring, when it is dug out and removed from the intestine, the chamomile preparation is still very moist and its substance has usually been completely transformed by micro-organisms.

Carlo does not find it easy to get hold of high-quality intestines. He knows a person at the slaughterhouse who is able to select the best cow for him. The animal should ideally be biodynamic, but at least organic, healthy, not fat and not treated with antibiotics. If there are antibiotics present in the cow there is a danger that no transformation process will take place inside the intestine. Carlo always uses fresh and even still warm intestines. He moistens the chamomile with the water from washing the intestine to make use of the bacteria existing there. The intestines are stuffed with chamomile flowers and these sausages are then buried in the first active 30 cm (12 in) of the soil.

Nettle preparation (504)

Nettles are harvested from the farm just before they flower. Carlo only uses the upper, softer parts of the plants and leaves (without the stem), and these are left to wilt in the shade. Carlo then places the wilted nettles into a terracotta pot. He puts in a layer of nettle, then adds water and repeats this process until the pot is full. The preparation is buried in the soil for a year after which it is dug out. Carlo first checks its quality – if it still smells of ammonia it means the process is not yet finished and Carlo leaves it in the soil a bit longer. When it is ready, the preparation should be half dry and, according to Carlo, have a very colloidal quality. This is why Carlo thinks it can be used for a longer time than horn manure.

Oak bark preparation (505)

The oaks (*Quercus robur*) for the preparation grow in the forest around the farm. Carlo rasps the crumbly outside layer of the bark with a cheese grater in order to obtain very fine material that can be filled by hand into a bovine skull. Carlo gets the skulls from the slaughterhouse. For him it is not so important whether they are fresh or not but they should not be very old. He uses about ten skulls each year and re-uses them for two to three years.

Dandelion preparation (506)

Carlo cultivates dandelions on his farm. He harvests the flowers in spring before they are completely opened. For the dandelion preparation, Carlo usually takes a dried mesentery with almost no fat. The mesentery is softened in water to make it supple. This water is subsequently used to moisten the dandelion flowers. A 'biological connection' is in this way established between the organ and the plant.

Carlo sews the filled mesenteries into balls that are not too big. He makes about twenty such balls each year. When the dandelion is unearthed, the form of the flower can still be seen, although inside it has been completely transformed. This is a criterion Carlo uses to judge the preparation.

Valerian preparation (507)

Valerian is grown on the farm. In Carlo's experience, the valerian preparation will only keep if the extract contains no chlorophyll. He therefore only uses the corolla and removes all green sepals. The petals are placed in a jar and water is added. He hangs the bottles up on a tree, and opens and shakes them each day. He continues to do this until a golden colour appears and the preparation has gained its full aroma. He then filters it, pours it into bottles and closes them with a cork stopper. The bottles with valerian preparation are stored in a dark place.

Applying the compost preparations

For the application to the compost heap, 2 g (1 tsp) of each preparation are placed into a small clump of Carlo's own humus rich soil and made into a ball. These soil balls are then inserted into the compost heap, one after another forming a row. Two millilitres (½ tsp) of valerian preparation are then diluted in 10 l (2½ ga) of water, and sprayed on the compost heap.

Burying and storing practice

Burying practice

Carlo explained that even though it was recommended in the Agriculture Course that some space should be left between them, Carlo does not leave much space between the pits containing the various preparations and yet he still has good results in obtaining the high-quality preparations.

Carlo has been burying the horns for producing the 500 in the same place for the last fifteen years. The other stuffed organs and nettle are also buried more or less in the same place every year. He commented that 'one should not fear doing things slightly different from the way Steiner recommended, after all he was not a farmer.' The stuffed organs for producing compost preparations are buried 20 cm (8 in) deep in the soil for half a year, in an area where the soil microbes are active. Nettle stays in the ground for about a year. He buries the skulls containing oak bark in a pit close to the house under the roof so that rain water from the roof can flow through them.

Storing the preparations

There is a special place on the farm, similar to a cellar and with a steady temperature, that has been specially adapted for storing the preparations. The 501 is stored in a glass jar in the sun and valerian in dark bottles. All the other preparations are stored in copper vessels and placed in a wooden box that is surrounded with peat. Carlo finds the copper containers very useful for storing the preparations since 'they retain moisture and because the copper does not oxidise in the darkness.'

Derived preparations and other applications
500 prepared (500P)

The 500P has been developed by Alex Podolinsky as a way of applying the compost preparations to the land. In Carlo's experience the 500P works well, even if there is no compost in the soil.

To produce the 500P, Carlo inserts the compost preparations into five separate holes made into the 500 contained in a box. The insertion of compost preparations is done as soon as possible after obtaining fresh 500. Carlo puts seven sets of compost preparations into each hole – equivalent to 14 g (½ oz) of each compost preparation. He then drops 14 ml (½ fl oz) of valerian into 1 l (1 quart) of water and sprays it over the 500 in the box. This first step is the warming phase, since by adding the compost preparations the 500 starts to warm up and the temperature rises to around 38–40°C (100–104°F). At this stage a reactivation of the biological processes appears to take place. After three months, the material is mixed together again and if the micro-biological process that leads to humus formation has not yet been established, the 500 is moved to another box and prepared again. After six months the 500P is ready. A new activity caused by the compost preparations has come about in the 500P and the structure of the substance becomes finer.

No immediate effect of applying the 500P can be observed: 'It is a longer process', Carlo says. An improvement in the soil and its structure gradually becomes visible. Carlo Noro thinks that the 500P is more complete in its effect than horn manure. We should, he thinks, be grateful to Alex Podolinsky for the opportunity of experiencing the effect of the compost preparations in the 500P, as by spraying the 500P the compost prearations are made effective in the soil.

Summary

Precision and diligence in making, storing and using the preparations are very important for Carlo and, in his view, essential for obtaining high quality preparations. He dedicates much time and care to observing the plant and animal ingredients in order to obtain the best possible quality ingredients. For example, he focuses on picking chamomile blossoms at precisely the right moment of opening. This attention to detail is present in the application of the preparations as well. Not only the water but also the 500 itself are warmed up to just under 37°C (99°F) before stirring, in order not to disturb microbiological activity. This approach requires Carlo to dedicate a large part of his working time throughout the year to preparation work. Carlo believes that this precision and dedication are necessary for obtaining high-quality preparations.

With regards to Carlo's approach to practical work, it stands out that he pays a lot of attention to the microbiological activity of the preparations and also of the soil. Special attention is given to supporting the microbiological processes taking place as the preparations are maturing while they are being stored and when they are applied.

In his experience, there is always scope for improving the quality of the preparations, and new possibilities for making improvements are continually emerging out of the practical work. Carlo feels it is important to be aware of what must be done at each particular moment and be awake to the exchange taking place between nature and humanity. For Carlo, the practical work generates inner experiences and understanding, and therefore the practicalities of life itself are more important than theory. With this in mind, Carlo's focus is to be recognised for the quality of his preparations and the results achieved using them.

Carlo believes that only by re-establishing the fertility of the soil and improving wild and agricultural biodiversity, can farm systems be strengthened and rebalanced so that diseases can be controlled in a natural way and food can be produced with high nutritional value. He believes that a balanced soil can be generated through the use of the preparations and this leads to balance above the soil as well.

9. Angela Hofmann, Sekem Initiative, Egypt

Dr Maja Kolar, Dr Reto Ingold

Introduction

Biodynamic agriculture is not very widespread in Africa, but in some countries like South Africa, Kenya, Namibia and Egypt there are biodynamic initiatives. With regards to preparation making, the Sekem initiative in Egypt stands out, since the biodynamic preparations have been made there for over thirty years and because the making of preparations has been adapted to both the desert environment and Islamic culture.

Sekem was founded in 1977 by Dr Ibrahim Abouleish with a vision to develop sustainable agriculture and a social community. Sekem's aim is to establish a blueprint for a fair and cultural society in the twenty-first century. Taking its name from the hieroglyphic transcription meaning 'vitality of the sun', Sekem was the first initiative to develop biodynamic farming methods in Egypt. It is located in the desert 60 km (37 miles) to the north east of Cairo, on the fringes of the Nile Delta close to the city of Belbeis.

Angela Hofmann is in charge of preparation making at Sekem. There, preparations are made for Sekem's own farm enterprises and for all the other biodynamic farms in Egypt. Angela is originally from Germany. She came to Egypt in 1981 and dedicated her life to the Sekem initiative as a biodynamic farmer. She is responsible today for the development of Sekem's new agricultural projects with, and for, the local farmers.

In order to promote sustainable agriculture and grow the raw materials needed by the Sekem companies, the Egyptian Biodynamic Association (EBDA) – a non-governmental, not-for-profit organisation – was founded. It provided training and a consultancy service for Egyptian farmers so that they could apply organic and biodynamic methods and gain the necessary certifications. So far, EBDA has facilitated the conversion of 140 farms and 2,500 ha (6,177 ac) of land to biodynamic farming (The Right Livelihood Award Foundation, 2003).

Egypt has an arid, semi-desert climate. It is characterised by hot, dry summers, moderate winters and very little rainfall (less than 80 mm, 3 in per year). Egypt's main source of water is the river Nile, which supplies over 95% of the country's water needs.

Dr Maja Kolar and Dr Reto Ingold visited Sekem between June 6 and 7, 2015, during the Members Assembly of Demeter International. On the evening of June 7, an in-depth interview with Angela Hofmann was conducted, and on June 8, a guided tour around the Sekem farms was led by Angela and her co-workers. This included a visit to the preparation house where there was an opportunity to ask detailed questions about the preparations. On the following day, Angela was interviewed a second time in order to answer the remaining questions regarding preparation practice.

> Sekem is made up of three closely interrelated entities: the Sekem Holding Company, comprising eight companies and multiple project-based initiatives, each of which is responsible for an aspect of Sekem's business value proposition; the Sekem Development Foundation (SDF), responsible for all cultural aspects, and the Cooperative of Sekem Employees (CSE), responsible for human resource development. Working together, they have created a modern corporation based on innovative agricultural products and a responsible attitude towards society and environmental sustainability (The Right Livelihood Award Foundation, 2003).

Farm portrait

The Sekem 'mother farm' was founded in 1977 in the region of Sharkia Governorate, which borders on the desert in the Nile Valley. It lies approximately 50 m (164 ft) above sea level and consists of 70 ha (173 ac) of desert land. The soils in the region are categorised as desert soils, or Aridisols (Soil Science Society of America, 2015). These soils are dry, receive less than 30 mm (1 in) of rain a year and organic matter is lacking.

The mother farm was started by Dr Ibrahim Abouleish on barren desert land. He took up the challenge of transforming desert land as part of his vision for creating employment and social development opportunities, while simultaneously improving the local environment.

The farm was started with the planting of a 30 m (100 ft) wide band of trees around the land. Some 50–60 Egyptian buffaloes were introduced and the production of compost started. At the same time, work with the biodynamic preparations began. Field and compost preparations were obtained from the consultant Georg Merckens. 500 and 501 were soon being sprayed on the fields and the process of enlivening the poor desert soil began. Later on, the buffaloes were joined by a flock of sheep and a dole of doves. The tall, white dove cots are a common feature of Egyptian scenery and were the first buildings to be erected in the area, apart from the stables. In his book, *Sekem: A Sustainable Community in the Egyptian Desert*, Dr Ibrahim Abouleish wrote that about three years into farm development the first signs of change became evident: 'The bushy, dark-green leaves of the trees were gradually starting to enliven the desert ground of the farm. I observed the number of insects and birds increasing on the farm, attracted by the trees and the treatment of the earth.'

Open stable on Sekem farm covered with palm leaves.

To start with, Dr Ibrahim Abouleish worked on his own to develop the farm with the help of his son Helmy. Soon he started looking for someone to care for the animals professionally. He found Angela Hofmann who joined the Sekem initiative in 1981. With her help, farming in Sekem began to really take off. Her main responsibility was to build up the dairy herd. In 1982, she imported forty Brown Swiss cows and a bull from different farms in Germany, because she found the Egyptian buffalo required too much care and attention, and the locally available Egyptian cows were not very productive. Angela Hofmann remembers how busy they were in those early days and how she could hardly sleep at night for worrying: 'In the very beginning we had to spend most of our time feeding the cows, because there was so little food in the desert. Nothing was growing and even if something was growing, the cows could eat up a whole hectare in one day.' So it took a lot of work to ensure a steady supply of fodder for the cows. Later on, in 2008, Holstein cows were introduced. Today there is a mix of breeds and around 127 milking cows and some 60 calves. About 500 indigenous fat-tail sheep are kept outdoors for meat production. There are eight dove cots, and more recently, 1,500 chickens have been re-introduced with the passing of the bird flu threat. There are beehives too.

A big effort went into reclaiming and cultivating the land. When treated according to biodynamic principles, and especially through the use of compost, the poor, light-coloured desert sands turned slowly into fertile soil. In 2007, Sekem successfully reclaimed some land in Minia, in the Sinai desert, and in Wahat el Baharya (three new Sekem desert farms). A second farm estate was developed there in order to grow agricultural crops. Medicinal and aromatic plants, and vegetables and fruits produced on the new desert farms provide the raw materials for processing by other Sekem enterprises. Today, the Sekem mother farm and processing facilities have grown to encompass a total of 270 ha (667 ac) of land on the farms of Belbeis, Wahat and the land in the Sinai.

Angela points out that Sekem is a growing ecological, social and cultural community in which 1,500 people work together, including gardeners, irrigation experts, livestock farmers, medicinal plant specialists, arable farmers, tractor drivers, technicians, compost makers, doctors, sales and marketing staff and teachers. There is also a kindergarten and a school with about five hundred pupils.

First steps in preparation practice

Angela Hofmann was born in 1958 in Germany. She grew up in the city of Stuttgart and already as a child wanted to become a farmer. She particularly liked cows and always looked for opportunities to spend time with cattle during her holidays. Her father was a Waldorf teacher and it was through him that she was introduced to anthroposophy and biodynamic agriculture. When she decided to take up farming it was clear to her that it could only be biodynamic agriculture.

Angela did an apprenticeship in Germany on biodynamic agriculture and household management. She was introduced to biodynamic preparation making during her apprenticeship on the family farm of Karl Tress, in Germany. There they followed the methods developed by Christian von

Wistinghausen. Angela then gained more experience with the preparations working together with the family Ackermann in Chiemsee, Germany, and on the Oswaldhof in Mattwil, Switzerland. During that time she sometimes joined preparation-making groups. She was, however, more familiar with using, rather than making, the preparations.

Angela's music teacher, Hedwig Kuch, heard about Sekem and told Angela's parents that Sekem would be a good place for Angela to go to work. When she first heard this idea, Angela was not particularly excited about it. Nevertheless, she made a visit to Egypt in autumn 1981 and it was soon clear to her that this was where she wanted to be. Angela only went back to Germany afterwards in order to organise her move to Egypt. Today she remembers the strong contrast between the mountain landscape of her family home and the desert, saying: 'After the green fields of my youth, where life is so easy, I ended up here in the empty desert and had to do everything by myself.'

Upon her arrival in Sekem, Angela was given the task of building up the animal herd and the dairy. She later took on responsibility for the whole farm and for making biodynamic preparations for Sekem and all its supplementary farms. As the number of farms using biodynamic methods steadily grew, the production of biodynamic preparations had also to increase. This now became Angela's main task. Angela also oversees field trials to research composting and the preparations.

Angela Hofmann and a co-worker.

How the work developed
Setting up local production of preparations in Sekem

When Angela came to Sekem, Georg Merckens, who was the first biodynamic consultant in Sekem, coordinated all the preparation work. To start with, Angela worked with him. She then gradually took on the lead role of preparation making in Sekem, introducing this work to the local people in Egypt as well. To this day, guiding and teaching others about the preparations is part of Angela's job.

In the beginning, Angela tried to follow the guidelines given by Christian von Wistinghausen and practised by Georg Merckens, and purchased the dried herbs from Germany. But she soon found that she could not simply follow the instructions step by step. A major challenge for Angela was how to adapt the making of the preparations to suit the special rhythms and conditions of the desert. One difficulty was in cultivating the preparation plants. It seemed almost impossible to grow valerian, dandelion and nettle. After many years, she managed to grow dandelion in the garden but for a long time it would not flower. In the end, though, she succeeded. Angela also spent many years trying in vain to introduce the nettle to this hot climate. She was only successful after she created a temperate microclimate using the shade of cover plants and water. Her attempts at getting valerian to flower have not yet been successful. She is convinced, however, that one day she will also find the key for introducing this plant to her garden.

Even if a dependency on a European supply of plant material could not yet be completely overcome, Angela did not search for substitute ingredients. She wants to keep to, and understand, the full potential of the original approach as it was given by Rudolf Steiner. Angela also explained that in Sekem they were glad to have a proven approach to follow with regards to preparation work: 'If you build up a farm and plant crops in the desert, there are so many new things to develop that we are grateful for being able to take plants that have proved their value so many times in other places. We also haven't been able to look into these questions in this beginning phase. If Rudolf Steiner had been here, I am sure he would have found other plants to substitute those that are difficult to grow here.'

Testing the effectiveness of the biodynamic preparations

Angela is very pragmatic when working with the preparations. Observing the effects of preparations is of core interest to her, while understanding how they work is not so important. To learn more about the effects of the preparations and to demonstrate their importance, Angela is supporting a variety of research projects on the farm. Apart from their scientific value Angela sees that: 'The results from the trials are important for our staff here since they can then experience the results directly for themselves. They weigh, count and measure the plants, and in so doing they see the benefits of the biodynamic method in the desert.' The farm trials also affirm the importance to Angela of 'making preparations with real enthusiasm, for doing so brings about good results and this, in turn, encourages us to make them even better next time.'

Trials comparing composts made using different organic components, such as vegetative plant waste, wood shavings and manure, have been carried out on the farm ever since Sekem started. The aim was to optimise the quality of compost produced under desert conditions. Good, humus rich compost, inoculated with compost preparations, has proven an essential ingredient for transforming desert sands into fertile soil. According to Angela, undertaking comparative trials was necessary, since there was no other research available on the production of biodynamic compost in desert conditions.

The work of Maria Thun was an important source of inspiration for Angela. Maria Thun came to Sekem many times to give lectures and carry out trials on cosmic rhythms and the efficacy of the preparations. Together they researched questions such as the effect of applying the preparations with the moon in different constellations. They found out that if the 501 is sprayed three times in succession when the moon is in a constellation of the same element (every fourth constellation), optimum results would be produced.

Angela also found evidence that the 500 and the 501 complement one another and need to be used one after the other. Better results are always found in terms of flowering, growth, and yield when both are used rather than just one alone. She also found that despite the hot desert sun, the 501 does not harm the plants if both preparations are applied during a plant's growth cycle.

The shift to moist preparations

One major change in Angela's approach to preparation practice has occurred only recently. When Helmy came back from a visit to Pierre and Vincent Masson in France in 2014, he was so full of enthusiasm for the work he saw there that Angela immediately arranged to go and visit them herself. She was impressed by their approach to preparation making, some of which came from Alex Podolinsky and some from the ideas developed by Pierre and Vincent Masson themselves. The aspect that most impressed her was the quality of the preparations and how they were stored in moist conditions. In her words: 'Because I knew from my experience in the desert how very important water and moisture is for life, I was immediately convinced that moist preparations must be better, even though it is very difficult to achieve this in the oppressively dry conditions of the desert.' She is convinced now that by keeping the preparations in a moist form, they are able to show the soil what it should become and how it should develop. Angela said: 'It works for me and we will also try to work this way in future here in Sekem too.' Angela is sure that this new approach is better, and despite the many years of storing preparations dry in the way she had been taught, she has now started to store them in a moist state. But it is early days and she is still exploring the best way to make changes in her practice so as to accommodate this new approach. As a result of the changes introduced so far, her preparations have changed from a state where the plant materials were still recognisable in the finished preparations, to a more completely transformed and colloidal state.

Angela observes how the preparations are developing in the store on a day to day basis and she has formed a special relationship with each single preparation. She observes how the 500 ripens in storage, develops a much finer structure and has a neutral smell. A similar process also happens with the compost preparations: they too are developing further in store. This ongoing transformation is quite a new experience for Angela.

There are many questions for her. How important is this development? It now seems to her that the preparations are not finished when they come out of the soil and that the process continues in the store.

With the introduction of moist preparations, Angela announced the start of a new series of trials in Sekem because she was convinced that, 'the moist preparations will give even better results.' Angela intends to continue working with the moist preparations in the future: 'We still need to adapt our way of working to these moist preparations because we are all so used to the dry ones. It is very difficult to keep them truly moist. But I am convinced that this is a better thing to do. So we are going to continue.'

Continuous learning

Angela explained that after thirty-three years of making and working with the preparations in Sekem, she still does not feel as if she knows much about them. She stated: 'I am still trying to understand the preparations. I do not know anyone who can say "I know everything about them"'.

Preparation making for Angela is not a rigid procedure. She has new experiences each year: 'Of course we always look forward to seeing how the preparations turn out. This is especially true in our climate where it can either be too dry or even too wet if there is too much irrigation water on the site where the preparations have been buried.'

As the person responsible for making preparations for all the biodynamic farms in Egypt, Angela is keen on developing the quality of the preparations ever further. She is continually looking for improvements, new techniques and new research results in this field. A lot of time and effort is needed to achieve good results, and this is also why Angela is now searching for a reliable person to take over the work with the livestock. Then she will be able to dedicate herself more fully to making preparations for Sekem and for the biodynamic farms of Egypt.

The social setting

Preparation work as a work to be shared

Work in Egypt is not so highly rationalised as in most western countries. The land work is always shared by a group. As farm manager, Angela has to deal with a lot of workers. She is surrounded by people from many different backgrounds – from illiterate farm workers to agronomists and other academics as well as people from the local authorities. All the farm workers in Sekem rely on Angela's guidance and advice. One can see that she has a special talent in being able, on the one hand, to give clear and simple instructions to the land workers who learn by doing, while on the other to present the background understanding clearly and patiently to the academic staff who require tangible scientific evidence. She therefore acts simultaneously as foreman for the one group and teacher and fellow researcher for the other.

In Angela's opinion, working with the preparations is something that should be shared with everyone on the farm, and not be a duty to be delegated to an expert. For her, work with the preparations is part of the regular farm work and everyone should be able to participate. Angela shares her knowledge and ideas about preparations with farm workers, interested academics and civil servants alike.

Angela explained what the preparation work with the farm staff looks like in practice: 'I need to organise myself and simultaneously organise them. Because of course it is strange work for these land workers. I really have to supervise everyone to get the preparation work done in the right way.' Much work and coordination is needed since 10,000 horns are filled each year. But once Angela has explained the work to the staff and they see the results on the fields, they come to 'love doing it and do so with enthusiasm,' Angela says.

One example that shows how Angela involves her team and works with them as one among equals, occurred when she was looking for a more efficient way of grinding the quartz needed for the 501. She wanted to get it as fine as that she had seen at the Massons' place. She discussed this question with her staff. One of them then found a person in Cairo with a ball mill who was able to mill the quartz into fine powder. Angela explained: 'I would never have succeeded without the input of my staff. Once my idea had been shared it was no problem searching the whole city to find the right person. Problems can be solved more easily through teamwork.'

Angela also arranges training and working seminars, she does consultancy work, supports new developments and encourages new ways of working together. She makes the land workers feel important and involved, and she trains them with a view of gradually handing over more and more responsibility to them.

It is remarkable that Angela was able to undertake this crucial role as a woman. It is not common in Egypt for a woman to be dictating the men's work, and yet she is very much respected. This can easily be observed accompanying Angela during a day with workers in the field and with students and teachers at Heliopolis University. Angela had to learn Arabic in order to communicate with the land workers and by doing so she was able to gain their respect.

Integration of preparation work with Islamic religious practices

Angela has always taken care that the work with the preparations fits in with local traditions and the guidelines of the Koran. This was only possible through the explanations of Dr Ibrahim Abouleish, who has a deep and modern understanding of the Koran. In his view the Koran clearly confirms the biodynamic approach and even invites the integration of this view of nature in the Islamic culture. Sura 55, ayas 5–9, of the Koran are often cited as evidence for this. Dr Ibrahim Abouleish has interpreted it as follows: 'Sun and moon follow their given paths, the stars and the trees bend their head in devotion, he (Allah) praised them to the skies and weighed everything. Man should not destroy this balance and maintain a precise relation between the nature realms.'

One example of the integration of preparation work with religious practice is Angela's advice to the farmers regarding the time for spraying the 501. She tells them: 'After you have done your morning prayer, you immediately start to stir horn silica, then you will be spraying at the right time in the early morning.'

Preparation practice

All the classical preparations suggested by Steiner are made and used on the Sekem farms and on all the farms throughout Egypt, which receive advice from the Egyptian Biodynamic Association. For a long time the compost preparations were imported from Germany, but since Angela started working with the preparations in 1982, she has gradually found ways of producing all the preparations on site, except for the oak bark and valerian preparations. In the 1990s Angela also made and used the Cow Pat Pit preparation from Maria Thun, but she found that the farmers were confused about how and when to use it, so she decided to stop using it. After many years of working with dry preparations, in 2014 Angela took up the challenge to keep the preparations moist in store, following the example of Pierre Masson.

Angela likes to make the field spray preparations in accordance with the Maria Thun Calendar, but because of the great number of horns (10,000 horns per year) being used, the farm crew is unable to carry out the work on the days recommended in the Maria Thun Calendar. The main preparation-making season always starts between September 10 and 20, for this is when autumn starts and when temperatures drop by 5 or 6°C (41 or 43°F).

Field spray preparations
Obtaining and handling horns

In Sekem, the horns from Egyptian buffaloes are used for preparation making. The horns of Egyptian buffaloes are bigger than those of the local Egyptian cows, they are very light and are available throughout the country. The horns from Egyptian cows are very small and would be inefficient for producing the 500 in the large quantities required.

Fresh horns from the slaughterhouse are washed and cleaned before they are filled with manure. Washing is repeated every year with the effect that they last longer. Because of the work involved, Angela is considering establishing some kind of horn washing festival in future.

The horns are not used for more than four or five years as they deteriorate quite quickly in the desert climate. Horns for the production of the 500 and the 501 are kept apart. Horns that have been once filled with manure are not used subsequently for producing the 501.

Horn manure preparation (500)

Angela and her co-workers collect the manure in the open run and the stable of the mother farm from their mixed herd of cows. They collect it regularly and use it as fresh as possible. In order to prevent too many fly larvae developing in the manure, a burn treatment is carried out on the stable floor once a month and it is treated with neem or cotton-seed oil. They do not give the cows any special feed during the period when they collect manure for the preparations. Four people work together with Angela to fill the horns of female buffaloes with manure. Angela explains: 'I have an agreement with the stable worker to collect all the nice cow pats every morning before they start to clean the open runs and store them in a big bucket. Every day, the preparation team gets a new quantity of good structured and fresh manure to fill the horns.'

Horn silica preparation (501)

Angela collects pure quartz sand in the desert. It is not pure white but has a sandy colour like the desert. Angela used to work with very fine riddled sand as recommended by Maria Thun. But since visiting Pierre Masson she has started to grind this sand very finely to obtain silica powder.

When filling the horns, Angela mixes the silica with water and makes a semi-liquid paste that can be poured into the horns. The filled horns are left standing for one day after which any water remaining on the surface is poured off. The horns are then sealed with dark clay before being buried. They are put 50 cm (20 in) deep into the soil, almost vertically with the opening upwards to prevent the contents falling out.

The 501 is made anew each year and if any good quality 501 is left from the previous year it is mixed together with the freshly ground quartz for making a new batch of the 501.

The 501 made with desert sand.

Stirring and applying the spray preparations

Angela originally showed the farmers how to stir the preparations. Today they also have a dedicated 'preparation engineer' at the Sekem farm who goes to the different farms and supervises the farmers as they stir and spray the field preparations.

Sekem farmers do all the stirring and spraying by hand as labour is still quite cheap in Egypt. Angela thinks that for 'the next thirty to fifty years no machines will be needed in Egypt'. The field spray preparations are stirred in a food grade 150–200 l (32–52 ga) plastic barrel.

The 500 is sprayed on cultivated fields while they are being irrigated with a sprinkler irrigation system. In this way 20 l (5 ga) of stirred 500 (containing 100 g (3½ oz) of horn manure) are applied per fedan (4,200 m^2, 1 ac). 500 is sprayed during the afternoon at least once a year.

The 501 is always stirred by hand in the morning. Two grams (½ tsp) of 501 per 10 l (2½ ga) of water are used. Farm workers start stirring after their morning prayers which are timed to coincide with sunrise, and spray it out with a mist sprayer directly afterwards.

The 501 is sprayed at least once during a crop's cultivation period, when the plants have got five or six true leaves, and is ideally sprayed three times, at times associated with the type of plant being grown according to Maria Thun's recommendations. For example, a leaf plant is sprayed when the moon is in the sign of Pisces, Cancer and Scorpio. Trials have shown that by spraying the 501 three times in succession in the same trine, the quality and quantity of yields is significantly increased. No further benefit was evident however by spraying more than three times.

Compost preparations

Yarrow preparation (502)

Yarrow is the preparation plant that grows most easily in Egypt. If there is enough water it grows like a weed. It is collected from the garden but because of the quantities required, it is not practical to follow the Maria Thun Calendar. When the yarrow flowers are at the right flowering stage, the plants are harvested by the farm workers. A group of special needs children and their teachers pick off the individual flower heads from the inflorescences. These are left to dry in a small dryer.

No stag bladders are available in Egypt. They are ordered from Germany, Slovenia or New Zealand, or are brought over from Europe by somebody visiting Sekem.

Angela always tries to make the yarrow preparation before June 24, before the summer solstice and before the plant sap goes down, in order to gather the maximum amount of precious plant content in the preparation. She soaks the bladders in yarrow tea to make them soft and moist, opens them and puts blossoms moistened with yarrow tea inside. She then ties up the bladders and hangs them under the roof of a house, making sure that they are safe from animals over the summer. She prepares 20–30 bladders each year. The bladders are buried in autumn inside clay pots.

Flowering yarrow in Sekem.

Angela explained that the yarrow preparation comes out with the flower structure still recognisable, but decomposition is completed during storage. She recounts: 'When we take the bladders out from the soil, the structure of the small blossoms is still visible, but after some time they are completely transformed. The decomposition process continues during storage because there are numerous microorganisms at work.'

A sample of Sekem's yarrow preparation.

Chamomile preparation (503)

Chamomile is one of Sekem's main primary products. It is processed into tea or sold as ingredient for pharmaceutical products. A lot of research has been done on this plant in Sekem with the aim of improving the variety and enhancing the taste of the dried herb. There are yearly big chamomile fields of several hectares on Sekem farm. Chamomile picking goes on over three months, starting in January and normally ending before April. The picking ladies pass through usually once a week. The blossoms are also picked by the famous 'Chamomile children'. This is a social project of Sekem that offers children who normally have to earn a living by child labour the opportunity to attend school. The flowers are picked 'when the blossom has a yellow tip, the lower part is half opened, the part above is closed and the petals are still white. This is the best moment to pick the herb for making tea and I think it is also the best stage for the preparation,' Angela explained. Angela can take the chamomile needed for preparation making, approximately 15 kg (33 lbs), out of the total quantity that is gathered on the farm.

Fresh cow intestines are always used for the chamomile preparation. If the intestines cannot be obtained from a cow from Sekem's own farm, it is not difficult to source it locally from small organic farmers, where the cows are fed on clover and hay. Ten kilograms (22 lbs) of intestines are needed per year to produce sufficient chamomile preparation for Sekem and the other supported farms in Egypt.

The preparation is made by wetting the dried chamomile flowers with chamomile tea and then stuffing it into the intestine by hand to form sausages. Two people work on one intestine. One person fills it from above and the other takes care that the intestine is well stuffed. The sausages are then put in a clay pot with holes, laid horizontally and buried in the soil.

The observation of three-week-old chamomile preparation revealed a sticky consistency and the presence of spring tails, which replace the non-existent worms in that hot climate. These soil organisms help to break down the larger particles and transform the preparation into a colloidal state.

Nettle preparation (504)

To Angela, having 'real nettle, *Urtica dioica*, grow in the desert is a miracle.' It is a plant Angela has struggled to grow for years. Angela receives nettle seeds from Germany, but the germination rates vary erratically: one year the seeds germinate well, another year they don't germinate at all.

The nettle needs a lot of nitrogen and water and only after Angela started growing it in the shade beneath the Moringa trees, did it start to grow. Angela said, jokingly, 'Where the nettle grows is our holy place in the garden.'

Nettle is collected when it starts to flower and the whole plant is used. The plant material is left for half a day in the shade to wilt. It is then put in a clay pot and buried for a year. A fresh sample was observed by the researchers. The plant material had been totally transformed during its time in the soil. Only the strongest stems were not fully decomposed. According to Angela 'the stems will disappear after some weeks if there is sufficient moisture in the store and, if conditions are right, the decomposition process will continue.'

Oak bark preparation (505)

Oak is not a native plant in Egypt and hence Angela buys in oak bark from abroad. Angela can order oak bark from the Wala gardens in Germany. She normally gets very finely ground bark of good quality.

There is no problem obtaining a fresh skull from an Egyptian slaughterhouse. Angela has overseen trials to compare the preparations made using the skulls of different animals: camels, Egyptian buffaloes, Holstein cows, sheep and goats. It was found in the trials that sheep heads were the cheapest solution. They can be obtained easily, especially around the time of Eid al-Adha, also called the Feast of Sacrifice, when nearly all Muslim families slaughter a sheep.

The brain is removed from the head using a piece of wire. The resulting cavity is filled with oak bark. The opening is closed with a piece of bone and coated with clay.

The filled skulls are put in a shady spot beside a water tank where water, continually leaking, creates a permanently wet hollow. Around 60–70 skulls are buried each year.

When this preparation is taken out from the soil it is 'very beautiful. It is raw and robust, soft and, at the same time, it is the strongest of all preparations' – this is how Angela described the result of its transformation in the soil. Angela's oak bark preparation has no bad smell and it decomposes to soft and dark mull, with a humus-preparation smell after being stored for one or two months. Angela is proud about her colloidal preparation: 'It is not easy to get oak bark in this highly decomposed state,' said Angela, 'and the biodynamic farmers here are very eager to have some.'

Dandelion preparation (506)

Angela is pleased that she has now succeeded in growing dandelions in Egypt. In the beginning, they had to care for and irrigate every single plant individually, otherwise it would not grow. After this difficult start, the plants started spreading on their own and now they grow in Angela's herb garden without any problems. No further planting is now needed. As only a small quantity is required, dandelion flowers are picked in winter by a group of special needs children who come to work in the herb garden between 8.00am and 10.00am each morning. They take their pickings to the dryer in the greenhouse each day. Dry dandelion flowers are stored in paper sacks.

Sections of fresh mesentery are easily obtained from the nearby abattoir. 'The butcher thinks we are eating them,' Angela chuckled. The mesenteries come mostly from cows of nearby organic farms. The mesenteries are always used fresh and there is usually very little fat attached.

The dried flowers are moistened with dandelion tea and wrapped into the mesentery to make small round packages that are tied up with cotton string. They make about eight packages each year. Two people normally work together: one person stuffs in the moist dandelion, the other makes sure the mesentery is held open until it is full. The dandelion packages are buried in the same way as the other preparations.

Valerian preparation (507)

Until now Angela has not been successful in growing *Valeriana officinalis* in Egypt. Valerian was planted several times and it even started to grow, but there have never been any flowers to harvest. Valerian extract is therefore bought from Hartmut Heilmann in Germany. Angela stores the bottles in a dark place.

Applying the compost preparations

In Sekem, a big effort is put into making good compost. All kinds of plant residues are added to the cow manure to produce compost. Temperature and CO_2 levels in the heaps are measured each day. The heaps are turned twice a week in the beginning, whenever the temperature rises above 65°C (150°F) or if the CO_2 level is more than 11%. The right amount of moisture needs to be maintained inside the heap. Care is taken to ensure the temperature of the compost heap rises to 65°C (150°F) in the first phase in order to kill off all the weeds, weed seeds and any possible pathogens. The compost then slowly cools down to below 50°C (122°F).

Only then are the compost preparations added. This follows research of 'Soil & More international' (2012) that showed that compost preparations work more effectively at lower temperatures. They are sometimes inserted into small earth balls, at other times they are put directly into holes in the compost. After inserting the compost preparations the whole compost heap is sprayed with the valerian preparation. Five millilitres (1 tsp) of preparation is diluted in 10 l (2½ ga) of water and stirred for 15 mins.

Because of the difficulty in keeping the compost heaps moist, the Sekem composting crew has developed a sophisticated programme to achieve this goal continuously through all seasons. By turning the compost heaps at least six to seven times, the compost can normally be finished within three months.

Burying and storing practice

Burying practice

Angela and her co-workers used to bury the preparations on different fields around Sekem farm. But since they often had problems with the irrigation – sometimes it was too wet, sometimes too dry – in 2014 they decided to make a preparation garden near the preparation store. The nettle and the assembled organs (except for the skull with oak bark) are now buried there in a small area making it easier to control irrigation and soil moisture levels. The compost preparations are only buried 10–20 cm (4–8 in) deep in the soil.

The field spray preparations are buried in the same place as the compost preparations. A large hole about 1 m (3 ft) in depth is prepared and the horns are put in, layer over layer. Each layer of horns is covered with fertile soil brought from another field. Horns filled with manure are buried horizontally; horns filled with silica powder are put in an upright position with the opening facing up.

The preparations are normally taken out of the soil in May when it starts getting hot. At this time the 501 is buried.

Angela Hofmann checks the moisture of the 500.

Storing the preparations

There is a special building on the farm dedicated to the preparations. In this building there is a bathtub which is used to store the large quantity of the 500 produced. The bathtub is embedded in a double-walled wooden box, containing peat between the walls. The compost preparations are stored in clay pots. The pots are kept in a wooden box with double walls filled in between with peat. The 501 is put in a large glass jar with a cloth lid and kept in a sunny corner of the preparation house.

Summary

For more than thirty years Angela has spent a lot of time integrating work with the preparations into the daily work of the farm in Sekem and other farms in Egypt. She considers biodynamic principles to be universally applicable as an approach to holistic agriculture and the production of healthy food. Together with her colleagues, Angela has managed to adjust preparation work to local climatic and agricultural conditions, and to integrate it within existing cultural and religious traditions and practices. Research trials play a major role in developing the work with biodynamic preparations and providing evidence of their effectiveness, thereby making them more socially accepted among various stakeholders.

Angela is glad to have been able to draw on the tried and tested experiences of using the European preparation plants and has found a way of growing most of them in the desert environment of Egypt. Neither oak nor valerian grow in the desert and they are imported from Europe. There are, however, ongoing efforts to grow these preparation plants in Egypt as well. For a long time it was not possible to grow dandelion or nettle, but Angela has already managed to create suitable environments for producing them.

Angela trained herself to share her knowledge and experience with Egyptian farmers and researchers, and she is very much valued and respected by them as an authority in the field of biodynamic agriculture. They trust her and are willing to follow her proposals. For Angela it is important that the preparation work is shared with the farm staff and fully endorsed by them. With this in mind, she does not tire to explain each step of preparation work in Arabic to her staff and to other farmers and land workers. Angela makes the preparations with real enthusiasm and she is able to encourage her co-workers to help her and develop this work further.

Angela and her preparation team have started to keep preparations moist while in storage since 2014. From Angela's point of view this is a big step forward from the dry preparations they were using in the past. She is convinced the moist preparations will give better results, as in this colloidal condition they are able to show the soil what it should become and how it should develop. She believes this is of particular importance in the desert environment where she works, she believes. A new series of research trials with the moist preparations is being planned in Sekem.

10. Andrea D'Angelo, Bairro Demetria Settlement, Brazil

Dr Ambra Sedlmayr, Dr Maja Kolar

Introduction

In Brazil, 3,000 ha (7,413 ac) of land spread over sixty three farms are under biodynamic cultivation. There are two main centres of work with the biodynamic preparations: Bairro Demetria in Botucatu and the farm Capão Alto das Criúvas of João Volkmann in Rio Grande do Sul. Preparations are also made by groups of farmers and cooperatives for their own use, but these normally only produce the field spray preparations.

Bairro Demetria is a rural settlement on the outskirts of Botucatu, in the district of São Paulo. It is mainly a housing estate, with the houses sparsely spread in a forest and connected by dirt tracks, forming what perhaps could be called an eco-village. This settlement grew out of the biodynamic farm Demetria that was established in 1974 by the Association Tobias, an association for furthering anthroposophical initiatives in Brazil. A number of anthroposophical initiatives, including a Waldorf school, the Associação Biodinâmica do Brazil (ABD – Biodynamic Association of Brazil) and the Instituto de Economia Associativa (ELO Institute) have been established there.

At the centre of the present case study is the preparation work of Andrea D'Angelo, who made herself available for participating in this study. Andrea D'Angelo is part of a group of about six individuals working with the preparations in Bairro Demetria, and the information on her views and practices is complemented with those of her colleagues. Andrea is part of the coordination group of the biodynamic training of ELO at Botucatu and is active as an advisor and teacher in the field of preparation work. Andrea has been influential in establishing preparation work at Bairro Demetria, both within the Biodynamic Association of Brazil and at the Demetria farm.

Dr Ambra Sedlmayr and Dr Maja Kolar visited Bairro Demetria on May 6 and 7, 2015. Interviews and conversations took place with Andrea

D'Angelo, René Piamonte, Deborah Castro, Pedro Jovchelevich and Paulo Cabrera. Guided tours of the Demetria farm and the farm of the Biodynamic Association were given and the researchers participated in work with the preparations: application of the 500 on Demetria farm, and making preparations with the students of the biodynamic agriculture training being run by the ELO Institute.

Farm portraits

The Bairro Demetria settlement is located some 750–800 m (2,460–2625 ft) above sea level on a gently sloping hillside. Average rainfall is 1,510 mm (60 in) per year. The climate is classified as hot temperate, with rain mainly in summer (333 mm, 13 in per month) and relatively dry winters (on average 137 mm, 5 in of rain per month). The average annual temperature is 21°C (70°F). The natural vegetation of the region is called cerrado – a savanna type ecosystem with shrubs and trees no taller than 12 m (40 ft). The area of Botucatu corresponds to a pocket of cerrado in a biome classified as Atlantic Forest (a subtropical moist forest biome). In Bairro Demetria, the preparations are made at the Demetria farm for its own use and at the ABD where they are produced for sale for biodynamic farms all over Brazil.

Demetria farm

The Demetria farm converted to biodynamic cultivation in 1974. The farm enterprise went bankrupt at the end of the 1990s and the farmers Paulo and Carolin Cabrera, who at the time were managing a neighbouring biodynamic farm (Sítio Bahia), were invited to take the farm on. They started working there in 2000 and had to rebuild the farm's degraded infrastructure.

Paulo and Carolin built the farm up around a herd of cattle, which is Paulo's passion. Maize used to be grown for silage to feed the herd of cattle through the winter months. However, Paulo realised that this led to a decline in soil fertility. He therefore started adapting the growing of irrigated tropical grasses (originally developed for conventional farmers) for the biodynamic system. He is very enthusiastic about using permanent crops to increase soil fertility and feeding fresh grass to the cattle all year round.

Sorghum field at Demetria farm.

In addition to keeping cattle and managing pastures, a number of other farm-related enterprises have been established, mainly under the direction of Carolin. These include a dairy processing unit, a bakery, a small jam and ice cream factory, and a shop selling natural products. Paulo and his colleagues also sell farm produce on three different markets in São Paulo twice a week. There are currently some twenty people permanently employed on the farm. Interns and volunteers are also integrated into the farm work.

At present the Demetria farm has about 42 ha (104 ac) of land. This is all that remains of a holding of more than 100 ha (247 ac) which Paulo and Carolin had been managing before their contract of tenure was changed. At the time of the visit, Paulo and Carolin were facing an impending threat that their rental contract would not be renewed for the coming year (2016) due to cash flow problems of the landowning organisation. Negotiations are ongoing.

The people involved in the preparation work on Demetria farm have been constantly changing. When Paulo and Carolin took on the farm, Paulo and Ronaldo Lempek made the preparations together. Ronaldo has since left the settlement and, being overloaded with other work, Paulo is unable to dedicate much time to this task. Various farm interns and volunteers, including Andrea, have been helping out. Andrea did the preparation work for Demetria farm voluntarily for three years. Currently she is working with Carolin, passing on her know-how so that Carolin can eventually take on the responsibility for the preparation work.

The Biodynamic Association of Brazil (ABD)

The Biodynamic Association of Brazil (ABD) was founded in 1984 on land belonging to the Association Tobias. The land rented by the ABD is currently divided into two plots. One plot is used directly by the ABD to produce the biodynamic preparation plants and grow seeds for its seed bank. Deborah Castro used to care for the preparation plants, now one of the employed gardeners is doing this work. The remainder of the land is let to a biodynamic grower who produces vegetables for a box scheme and a biodynamic stall on the market in São Paulo.

Since 1988, the ELO Institute has been running a biodynamic training course. Andrea D'Angelo is part of the coordination team and guides the students in the study of the Agriculture Course. René Piamonte and Deborah Castro teach a practical module on preparation making.

Andrea D'Angelo.

René is an international biodynamic consultant who trained at the Dottenfelderhof (a famous Demeter farm and biodynamic training centre in Germany) and then came to Brazil, where she carried out the preparation work for the ABD for many years. René learned how to make the preparations from Marco Hofmann and Christian von Wistinghausen. René is the main teacher of preparation making in Latin America and this, he argued 'is the reason why there are no significantly different approaches to making biodynamic preparations in Latin America.'

Deborah Castro was employed by the ABD from 2000 until early 2015 in order to produce biodynamic preparations for sale. She was also in charge of demonstrating the practical making of the preparations to students on the biodynamic training courses. Her contract was terminated at the beginning of 2015, however, due to financial insecurity. Deborah was taught how to make the preparations by Andrea D'Angelo and René Piamonte. Because she also understands French, she picked up some books on biodynamics written by French authors, such as Pierre Masson. This led her to change some of her practices and, in particular, led her to store preparations in a moist state.

First steps in preparation practice

Andrea was born in 1976, the daughter of a conventional farmer. She grew up in the city of São Paulo. Already in her teens she developed an interest in environmental conservation. She wanted to learn about nature and how humans can work with it, and went on to study agricultural engineering in Piracicaba. The chemical approach to agriculture that she learned about seemed too aggressive and she soon started seeking for an alternative approach. She also remembered wanting to 'work with the spirit of nature rather than being purely production focused and materialistic.'

During her first year at university there was a professor who gave inspiring lectures about soil fertility that were based on his understanding of biodynamic agriculture. This inspired Andrea because 'this was an approach very close to the living world' and made her enthusiastic about the possibility of a new form of agriculture, namely biodynamic agriculture. Andrea read the Agriculture Course with great excitement. She recounted 'when I met biodynamic agriculture I fell totally in love with it. There I found something that is really truth ... something that the world needs and that people need.' Soon afterwards she became part of the team organising the Brazilian biodynamic conference in 1998.

During this biodynamic conference at the University of São Paulo in Piracicaba in 1998, Andrea listened to, and met with, a number of well known biodynamic researchers from Europe. She decided that she would like to go to Europe, to get to know the places these people came from and learn more.

In 1998, during her last year at university, Andrea attended the biodynamic training course run by the ELO Institute in Botucatu. She was particularly impressed with the module on the biodynamic preparations, taught by René Piamonte. She remembers it as being a turning point in her life. It was the moment she connected herself emotionally to the work with the preparations and realised 'that this is something for my life, it has something to do with me.' What touched her most during these lessons was the experience of the 'extreme aliveness' of the preparations. She felt passionate and excited about using simple plant and animal parts in order to create something that is really alive and can be used in agriculture in a way that is fruitful for the Earth. Another aspect that moved her was that work with the preparations needs to follow the rhythms of nature. It is a way of integrating one's work into the greater rhythms of nature and the cosmos. This aspect of preparation work is both important and deeply meaningful to her.

Her enthusiasm led her to take up the ABD's preparation work, which, since the departure of René Piamonte to Argentina three years previously, had been left largely unattended. She already started working for the ABD on a part-time basis during her last year at university. She started by tidying up the storage box and sorting out the preparations. The ABD team supported her work and when Andrea finished her studies she went straight on to do an internship with the ABD. Her intention was to renew its preparation work and thereby support the Association in its efforts to further the development of biodynamic practices and provide assistance to farmers. Within a year, Andrea had started growing most of the preparation plants and was making preparations for use in biodynamic projects and by consultants.

A year later, Andrea's long-standing desire to visit Europe and learn more about preparations, made her look for someone who could take on the work she was doing with the ABD while she was away. She found Deborah Castro, who was studying agronomy and doing an internship at the ABD and who was happy to take this work on, starting in the year 2000.

While she was in Europe, Andrea visited a number of farms and biodynamic initiatives. This included a season working with Christian von Wistinghausen. From the beginning of 2001 until 2003, she conducted a research project about the natural environment of the preparation plants at the Natural Science Section in Dornach, mentored by Jochen Bockemühl and Hans-Christian Zehnter. During her time in Dornach, Andrea also had opportunities for intensively studying the Agriculture Course. At the end of her project she moved to the Dottenfelderhof, in Germany, in order to learn about, and have contact with, animals. While she was there, she worked with the cattle and assisted Knud Brandau who was then in charge of the preparations. Soon after returning to Brazil she started teaching the Agriculture Course on the biodynamic training course in Botucatu. She joined the team coordinating the biodynamic training and is still involved with it today. Andrea is very keen to exchange experiences and share ideas with other people working in a similar direction, like René Piamonte, Deborah Castro or Paulo Cabrera.

Since returning from Europe, Andrea has not yet found a way of being paid for her preparation work. She is, however, continuing to work voluntarily and offers her help to anyone needing support and advice. A new cycle of project funding would mean that she could devote herself to the work she feels really passionate about – supporting and teaching farmers how to work with the preparations themselves.

How the work developed

The development of Andrea's practice and understanding of the preparations has been marked by her search for practical learning experiences. Her time in Europe was entirely dedicated to deepening her understanding of the preparations by connecting with those with expertise and by doing practical research.

New insights into the preparations came to Andrea mainly through her own practical work with them and when working with farmers or students. Questions, observations and different approaches provided new insights into the preparations and helped to advance her practical skills.

One such practical learning experience concerned the question as to whether a fresh skull or an old skull should be used for the oak bark preparation. Andrea had learned to use a fresh skull and fill the oak bark into the still intact meninges (brain skin). She then had opportunities for joining preparation-making events in which old skulls were used and where the oak bark was filled directly into the bone cavity of the skull. This experience enabled her to contrast and compare the two different methods. She concluded, like René Piamonte, that although she could not observe a difference in the final oak bark preparation, the meninges fulfils an important function. She sees it as being a sensitive skin, like that of the intestine, mesentery or bladder. It adds a living element to the process of transforming the oak bark and is connected specifically to the nerve-sense system. When using the skull on its own, this sensitive skin, and the life forces related to it, are missing, and the bones of the skull act 'only as a vessel'.

While she was in Europe, Andrea came across the idea of storing the preparations in a moist, colloidal state. This was new for her, but it was not a method that particularly attracted her interest. As regards the 500P, Andrea feels that it is very mineralised and too advanced in its decomposition. She finds that 'whereas 500 still has a lot of potential within it, 500P is already "finished".' Andrea, however, remains open to the possibility that the moist storage of the preparations is better in certain situations or is preferred by some people. She feels she would need to carry out further research in order to come to a conclusive judgement.

In both these examples, having observed and evaluated alternative approaches, Andrea returned to the methods she had originally learnt.

Andrea D'Angelo's understanding of the preparations

A new quality emerging from preparation work

Preparation work for Andrea is 'a new way of dealing with the substances of nature', and a possibility for 'developing a new attitude towards the beings of nature and to spiritual beings'. Andrea is fascinated by the new science expressed in preparation making. It contains a wisdom that most people are not yet able to fully grasp, but also a truth that can be felt and experienced as deeply meaningful. The applied wisdom of the preparations results in a new consciousness. It brings 'a new quality of nutrition, for human beings and for the earth.' This new quality is described as altruistic consciousness, a capacity that is to evolve in the future but which can already appear today through the preparation work. Andrea explained: 'This capacity allows one to almost see a new world arising ... it seems that some young people already have this organ to perceive a special type of non-material quality.' Andrea explained how this new quality, brought about through working with the preparations, is not unlike the quality of wakefulness brought about by regular meditation – preparation work can be likened to meditation.

Preparation work as a meditative practice

Andrea believes that working with the preparations is a form of meditation that is related to the will. She explained that it is an activity carried out in freedom and that a strong effort of will is needed to actually do the work. Andrea said;, 'One needs to find the strength within oneself to do it, to find the ego power, just like in a meditation. The individual has to make the inner decision: "I want to do this".' Since farmers tend to be very busy, a lot of willpower and focused intention is needed to do this extra work. Intention and a strong will results in good preparations being made and serves to generate that 'altruistic quality' previously described. Andrea believes that developing such conscious intentionality is crucial to the quality and effectiveness of the preparations. She has noticed that the preparations are of a much higher quality when the person making them is fully present.

Andrea also likens preparation work to that of meditation. She feels that the preparations should be in the consciousness of farmers and that they should try and work with them every day. Throughout the year there are many tasks connected with the preparations that can be done to keep the preparations in one's consciousness like a daily meditation.

Andrea described her meditative approach to the preparations as follows: 'My way of dealing with the preparations is … I don't sing or recite poetry … I respect those who do, but it's not so important to me. My attitude is to be totally present. I try to do it as a meditation and do it every day. I think it has something to do with willpower, with being fully present and not thinking of something else.'

The universal nature of preparation plants

Ever since Andrea began teaching biodynamics, she has been repeatedly asked whether some of the preparation plants should be substituted by tropical plants. This is not a question Andrea asks herself. Her understanding is that the plant world is universal in character and that, unlike the animal world, which is more strongly differentiated by regional forces and conditions, the entire plant world of the earth forms a single whole.

Andrea feels that there is no real need to find substitutes for the European preparation plants, since most of them can also be grown in the tropics or brought in from nearby regions. In fact, she sees a danger in seeking substitutes without having the necessary spiritual faculties or insights. She recommends that people should always try to understand the classical preparations first.

Her work in Dornach with Jochen Bockemühl, Hans-Christian Zehnter, Manfred Klett, Christian von Wistinghausen, Peter Blaser and others, led her to recognise that working with the preparations is of universal human value and has meaning for the earth as a whole. 'I don't only work with the preparation to benefit this particular place here, where I am living, when I do it here it has a meaning for the entire earth.' This backs up her belief that the same preparation plants can be used all over the globe.

Assessing the quality of the preparations

In Andrea's experience, the quality of the preparations largely depends on the quality of the original ingredients – of the plants and the animal organs. If the mesentery is too fatty, or the dandelion or chamomile mouldy or going to seed, the preparations will necessarily be of lower quality.

When she digs up the preparations, Andrea assesses their quality by first checking how much earth is mixed in with them. Although she feels that a little bit of soil mixed in with the preparations is not a real problem, when she takes them out herself she tries to reduce the amount of soil as much as possible. Andrea also checks the smell – the preparation should not stink – and there may still be something of the flower quality present. She also uses touch to assess the quality of the preparation.

The most important qualitative feature for Andrea, however, is that the preparations retain some of the structure of the original plants, rather than being fully decomposed. In her opinion, if the structure of the plant can still be recognised, more of the plant's being is present than if it has fully decomposed. Andrea feels that if the plant is fully decomposed it is already becoming soil or compost, and that the process of transformation has gone too far. She believes that it is important to still recognise the preparation she holds in her hand and that each one has its own specific quality and appearance. Dry, or relatively dry, storage of the preparations helps to maintain the plant structure.

The effects of the preparations

Andrea finds it difficult to show any specific effects of the preparations, because she is not able to accompany them on a farm in detail throughout the year. When applying the preparations to the compost, however, Andrea feels that a change takes place: the compost becomes fresher and is more alive. In her experience, prepared composts tend not to be so heavy or smelly. As regards horn manure, she feels that the atmosphere of the place changes when it is applied – it is enlivened and has a more harmonious mood. She also thinks that the 500 changes the soil, making it more alive. She has observed how the 501 influences the developing shape of onions and bananas, or the smell of chamomile. Her general perception is that the 501 changes the light quality of a field, bringing to it some form of etheric or spiritual light.

Futher aspects of Andrea D'Angelo's approach to the preparations

René Piamonte has shared some aspects of his understanding of the preparations. These are accepted by Andrea D'Angelo and many other preparation makers in Latin America who have learned from René.

The organ functions continue

René's understanding is that the original functions of the animal organs used to contain the preparations continue to work during the process of maturation. From his point of view, this is key to understanding how an organ should be used in order to produce the preparation. For example, the function of cow horns is to concentrate forces, and they will continue to do this in the soil. When they are beneath the soil in winter they concentrate the forces of the 'earth's I – (o eu da terra)' inside the horn. By this same logic, René finds the idea of hanging up filled intestines and mesenteries in the sun, to be counterproductive; these are internal organs that 'do not need to receive anything from the environment', but instead work inwardly to effect what is inside them. From his point of view, the bladder is different, as it is an organ which is sensitive to the outer environment.

Resilience of the farm individuality

René said that nature is continually being weakened by 'the attack of civilisation on nature' and that it requires more than organic techniques to counteract this attack. The biodynamic preparations are 'spiritual impulses, incarnated as processes.' They bring spiritual impulses to the earth. His understanding is that biodynamic agriculture helps a farm to approach the archetype of a farm individuality, and that the preparations are the means whereby this archetype can incarnate. This, from his point of view, is the secret behind the high degree of resilience found in biodynamic farms. Just as a person in touch with their inner being can bounce back again after receiving a shock, a farm that is close to being a farm individuality is also better able to cope with stress and shock.

Preparations as guides for inner development

Deborah and René both pointed out the importance of the preparations for personal development. The experiences and questions that arise when working with the preparations lead to an ever deeper study and understanding of anthroposophy. This in turn has a transformative effect on the person involved. This is one reason why all biodynamic farmers are encouraged to be involved in making preparations themselves. They can then discover the relevant questions and experiences that can help them deepen their biodynamic practice. Working with the preparations in this way is instrumental in cultivating an approach to life that informs good biodynamic practice.

The social setting

At the Bairro Demetria, making the preparations always involves a number of people. At the ABD it has so far been mainly Deborah Castro who has organised the work and made most of the preparations. She also demonstrates how the preparations are made as part of the biodynamic preparations module of the ELO biodynamic training course. At Demetria farm the making of preparations is done by different people; those involved change from year to year, and even in one year it is possible that more than one person is making the same or different preparations with no coordination between them. The application of preparations at Demetria farm often involves the farmer, volunteers and neighbours.

Andrea rarely ever makes the preparations on her own, except for her own garden. She always does it with other people, farmers and students. She believes that it is important to be open about the preparations and that more people need to understand how biodynamic agriculture works. To this end she always invites interested customers and neighbours of the Bairro Demetria to join in with whatever preparation work needs doing.

Preparation practice

The eight classical preparations suggested by Rudolf Steiner are all made, as is the CPP, known in Brazil as '*preparado fladen*'.

At Bairro Demetria, the constellation of the moon does not play a role when deciding which day to make or use the preparations. Attention is instead focused on whether the weather is suitable for carrying out the work (for example, sunny days for making the 501 preparation). The moon phase is sometimes considered. René Piamonte explained that the power of the preparations is more powerful than the influence of the moon.

Field spray preparations
Obtaining and handling horns

Conventional horns are sourced from abattoirs. Sometimes, people who know that Demetria needs cow horns keep horns for them. The horns can be used three to four times. When they get very thin and crumbly they need replacing.

Horn manure preparation (500)

Fresh manure is collected directly from the pasture or the milking parlour on the morning of the preparation-making day that takes place around Easter. The horns are filled using wooden spatulas or by hand. To ensure the manure is filled to the very end of the horn without any air pockets, the horns are hit on the concrete floor. The ABD uses 1,000 horns per year for making the 500.

At Demetria farm each field is sprayed individually. For stirring, Andrea uses about 100 g (3½ oz) of the 500 per hectare (2½ ac). This is stirred in a barrel with about 170 l (44 ga) of water. In the last 20 mins of stirring, CPP is often added. 500 is applied to the fields in the afternoon or evening. For this, a number of volunteers join the work. Buckets of stirred preparation 500 are taken into the field and branches from trees and shrubs near the fields are used as brushes to sprinkle the preparation on the soil.

Stuffing horns for the 500.

Horn silica preparation (501)

Brazilian quartz crystals are used. They are ground by hand, using something akin to an iron pestle and mortar. Iron splinters are removed using a magnet covered in toilet paper, to facilitate the removal of iron particles afterwards (a practice taken on from João Volkmann). The quartz thus obtained is then ground further using a normal pestle and mortar and sieved through a very fine sieve (nylon tights). The resulting powder is mixed with water and filled into horns that have not been used for making preparations before. The horns are placed in sand with open ends up for one or two days so that the excess water accumulating on top can be poured away. The opening of the horn is then closed off with a piece of clay. It is buried in the ground in a sunny spot about 40 cm (16 in) deep around Michaelmas. It is lifted at the end of March or beginning of April – the end of the Brazilian summer. 501 is stored in glass jars on a window sill. It is applied at least once or twice a year, depending on the light quality experienced in a given field, and on the crop being cultivated. Spraying is done early in the morning to ensure the work is finished before 10.00 or 11.00am. It is sprayed as a mist over the fields.

Stirring

Andrea has not stirred on her own for a long time. She always does it with students, neighbours and children. Andrea shows them how to stir and they then take turns to stir. She tries to create a joyful and meditative mood. It is important to her that attention is focused on the work in hand and that there is an awareness of the farm and its people. While they are stirring, Andrea or the farmer Paulo will often explain something about the preparations or answer questions.

Compost preparations

The animal material is usually obtained from a regional abattoir. A person familiar with the requirements saves the organs for the Bairro Demetria group. Sometimes it is possible to obtain the organs from a cow slaughtered at home for a special occasion by someone in the area (like for a wedding).

> **Personal experience of preparation work at Demetria farm**
> Dr Ambra Sedlmayr
>
> Dr Maja Kolar and I were coming from a visit to a place at Demetria farm, when René Piamonte and Paulo Cabrera, along with some of their colleagues, crossed our path. They were carrying a barrel that turned out to contain ready stirred 500 to which CPP had been added in the last 20 mins of stirring. The barrel was placed on the back of a tractor and driven down to a field in the valley. A group of people, including some school children, followed the tractor to help with spraying. They seemed happy, excited and motivated to help with this work, which they had never done before. At the field, each person was given a bucket filled with preparation and people started to rip branches from the shrubs and trees bordering the field. These were dipped into the buckets and then used to sprinkle the preparation on the field. Some six people were walking in a row, side by side, at a distance of some 5–6 m (16–20 ft), applying the preparation as they walked down the field. Only René Piamonte refrained from using branches, since he had his own special spraying technique: with a small plastic container he would take a sip of preparation out of his bucket and fling it energetically over the field, so that small droplets would be produced. Preparation spraying created a joyful and peaceful mood, very much in harmony with the ending day. As we were spraying, night approached. The skies turned an orange red. Somehow, Maja and I missed jumping on the tractor before it departed back to the farmyard, and we had to find our way back on foot through dark and unknown forest paths.

Yarrow preparation (502)

In Brazil, the hunting of deer is prohibited because this species is facing extinction. Bladders must therefore be imported. Andrea gets the stag bladders from the Mäusdorf Preparation Centre in Germany.

The yarrow is grown by the ABD. The yarrow flower heads are cut with scissors or by hand when they are well into flowering. Yarrow flowers develop very slowly and there is no rush to pick them. If the stalk or pedicels are too hard they are removed; if they are soft enough they can remain. Harvesting is done from October to February on Flower days. The yarrow is dried and stored in cardboard boxes until required. To stuff the bladder, both the dried bladder and the dried yarrow are moistened with yarrow tea. The filled bladder is then hung up in a sunny place throughout the summer. In autumn it is taken down and buried.

The bladder in the soil is surrounded with wood on the sides and on top, so that it can be found more readily in spring.

Andrea D'Angelo shows the 502 at Demetria farm.

Chamomile preparation (503)

Chamomile is grown in the gardens of the ABD and at the Demetria farm. Fresh flowers are picked each morning between 8.00am and 10.00am for around twenty days in August. The flowers are dried in a solar drier and stored in cardboard boxes until Easter when they are filled into the intestines. Fresh intestines are flushed out with water, filled with air and hung up to dry until the preparation is to be made. At the ABD, the chamomile flowers (along with some stalks that have been harvested together with them) are moistened in hot water. The dried intestine is placed in warm water until it becomes supple. Andrea prefers to use chamomile flowers without any stalk, and to work with fresh intestines that have been emptied but not flushed out with water, so as to preserve the inner membrane of the intestine as much as possible. If dried intestines are used, she uses warm water or chamomile tea to moisten and enliven them. Chamomile is the preparation which shows the least amount of plant structure when it comes out of the ground. Some soil is usually mixed up with the chamomile, since the intestine is so thin that it is very difficult to keep it separate.

Nettle preparation (504)

Nettles are specially grown to produce the preparation. They have to be shaded and develop into soft plants with many leaves and have relatively short inter-nodes. Flowering starts at the end of November or beginning of

December, which is the ideal time to make the preparation. After mowing it, the nettles are left to wilt for two or three hours. The whole plants are then placed in a pit in the ground that has been lined with wood on each side. A tile is placed on top of the nettles and the pit is covered with soil. The physical barriers were developed because, without them, it is not possible to recover much of the nettle preparation from the soil. Another method that is sometimes used, is to place the nettle plants into a loose cotton or jute bag and then bury it. One whole year later, the bag with the finished nettle preparation inside is taken out. Sometimes clay pots are also used for burying the nettles. When it comes out of the ground the preparation is often mixed with soil – an accepted reality. Despite nettle stalks having also been included, they are quite soft and very little is noticeable in the finished preparation.

Oak bark preparation (505)

The use of a fresh cow skull is preferred. The brain is washed out using water pressure.

The oak bark comes mainly from João Volkmann in the south of Brazil because the oak trees planted in Bairro Demetria are still too small to harvest their bark. If the oak bark is dry, it is left to soak in water overnight before being finely grated with a cheese grater. This practice was introduced by Luiz Felipe Ricca (who trained with Carlo Noro in Italy) to prevent damage to hands while grinding (because the moist bark is softer).

In autumn, the oak bark is filled into the skull with the meninges intact. The presence of the meninges is deemed very important in the production of a quality preparation. There is a pond in the Cerrado forest near to the ABD, where the filled heads are buried.

Students watch the burying of the 505 in a pond in the Cerrado forest.

Dandelion preparation (506)

Dandelion is also grown at the ABD. For harvesting, the flower heads need to be open on the outside and have closed centres. Dandelions are dried in the sun or in a solar drier so as to remove moisture quickly and halt the process leading to seed development. The dandelion flowers must stay dry and not get mouldy. Moistened dandelion flowers are placed into a piece of greater omentum and good-sized packets are assembled by tying the packets with a cotton string.

Andrea likes to see the dandelion preparation with a structure in which the flower remnants are soft and not hardened. The quality of the dandelion preparation is affected by the greater omentum: if it has too much fat, its quality is negatively affected.

Deborah Castro demonstrates assembling dandelion packets.

Valerian preparation (507)

So far the valerian preparation has had to be purchased from the Mäusdorf Preparation Centre, because nearly all attempts at encouraging valerian to flower in Brazil have so far failed; only rarely is a flower produced. If a plant flowers, its seeds are harvested and used to produce the next generation. The aim is to develop a new variety that will eventually make it possible to produce valerian preparation from locally grown plants.

Applying the compost preparations

The consultants from Botucatu encourage farmers to make compost with both animal manure and straw or green matter. After building the pile, the five compost preparations are introduced. Normally each preparation (about 5 g, 1 tsp) is rolled into a small, clay ball and then added to the compost heap. The most important consideration with regard to the order of inserting the compost preparation is to have the nettle preparation in the centre of the pile and between the other preparations. Five to ten drops of valerian preparation are stirred in 20 l (5 ga) of warm water for 20 mins, before being sprayed over the whole pile. Finally, the compost heap is covered with straw. When it is turned, the preparations are added again, in the same way as previously described.

A liquid fertiliser is produced using water, cow manure, plant materials, compost or soil, ashes and sometimes even bakers' yeast. This mixture is stirred until it is homogeneous. The clay balls containing the preparations are added to the mixture and valerian is sprayed on top.

Burying and storing practice

Burying practice

The assembled preparations are buried some 40–60 cm (16–24 in) deep, depending on soil quality. Places are chosen where there are not too many roots that could grow into the preparations. The different preparations are buried in holes at least 5–10 m (16–33 ft) away from each other. When working as a consultant, Andrea likes to suggest that the preparations are buried near to where the farmers live so that the preparations are held more easily in their consciousness. Andrea decides where the preparations are to be buried according to the feeling she has about the particular farm and the level of awareness for the preparations among the people working on the farm.

Andrea likes to make and bury the preparations in early autumn. She feels that it is important for the preparations to be buried around Easter time and that they have a long period underground so that the transformation process is fully completed. In her opinion, the preparations should ideally stay in the soil from April to September.

Storing the preparations

At Demetria farm there is a wooden cabin in which the preparations are stored, at the ABD there is a store room in which everything connected with the preparations is kept as well as the preparations themselves.

The preparations are stored in unglazed clay pots that are sunk into a mixture of bark, black soil and xaxim. The preparations are loose, have a low level of moisture and the structure of the preparation plants is easily recognisable. It is important in this part of the world to keep a close eye on the preparations in storage to prevent moths settling and moulds appearing.

> ## *Teaching preparation making*
> Dr Maja Kolar
>
> Deborah Castro, who is in charge of the demonstration of preparation making for the biodynamic training of ELO, had carefully prepared all the materials needed for making the preparations. Some tables had been set up in the garden of ABD in front of the preparation storage room. This was a peaceful place, surrounded by trees and plants and lots of birds. A group of around 20 participants, of mixed gender and age, and from different parts of South America, gathered there. For most of them, it was the first time they were faced with the practical work of making the preparations. Deborah therefore explained all of the steps while she was demonstrating them with the help of students, who followed her with great interest. One could feel the motivation of the students as they could hardly wait to start the work. The tasks were shared and students helped energetically and without bias, whether it was with washing and filling the skulls, intestines or horns. Students also participated in burying the preparations. The preparation materials were handled very respectfully and carefully. René Piamonte was present throughout, mainly answering questions and giving some theoretical background on the work at hand.

Derived preparations and other applications
Cow Pat Pit preparation (CPP)
Cow Pat Pit, or barrel, preparation developed by Maria Thun, is produced for the ABD by Deborah Castro. Manure, basalt and eggshells are mixed for one hour. A hole in the ground is lined along the sides with wood and on the bottom with bamboo so that it looks like half a barrel buried in the ground. The bamboo on the bottom is important to make sure the preparation stays in the barrel and is not absorbed into the surrounding soil by soil organisms. The 'barrel' is then covered with wood and straw. Cow Pat Pit preparation is produced throughout the year. Green manures are often used in the tropics because it is important never to leave the soil

uncovered and exposed to the strong sun and rain. When green manure is cut, Cow Pat Pit preparation is applied on the material in order to guide the decomposition process. One method of applying the Cow Pat Pit preparation is to stir it in during the last 20 mins of stirring horn manure. Two hundred grams (7 oz) of CPP are used per hectare (2.5 ac).

Summary

At the Bairro Demetria there are several people involved with the preparation work: Andrea D'Angelo, Deborah Castro, Paulo and Carolin Cabrera and René Piamonte. Andrea D'Angelo took on the preparation work in 1999 for the ABD after René Piamonte had left. She helped at various times to further work with the preparations at the Bairro Demetria and has dedicated a lot of time and effort to studying the preparations.

The practical experience of working with the preparations is what proved most fruitful to Andrea as she developed her interest in, and understanding for, the preparations. It was the hands-on preparation-making session with René Piamonte that first fired her impulse to work with the preparations. Later, she sought practical learning experiences in Europe and joined preparation makers and researchers in their work. For Andrea, it is the quality of aliveness she experiences when working with the preparations that convinces her that biodynamic agriculture is the agriculture of the future.

Andrea's preparation work is characterised by a strong focus on the practical and the meditative aspects of the work. She feels that an altruistic consciousness can develop through working with the preparations leading to a new way of dealing with nature and food. The consultants at Bairro Demetria are in agreement amongst themselves that the preparations are a guide for inner development, and that working with them strengthens the link with the rhythms of nature and awakens interest in gaining a deepened understanding of biodynamics and anthroposophy. It opens up a path of learning and a new way of thinking and self transformation. This is why, at the ABD, they seek to encourage farmers to make the preparations themselves, so that their understanding of biodynamics is deepened.

For Andrea, retaining the original structure of the plant material in the preparations is an important reflection of their quality. The guidelines given by Christian von Wistinghausen and Jochen Bockemühl are generally followed. It is understood that the plant world has a universal character and that there is no urgent need to find tropical substitutes for the European preparation plants. Andrea believes that the preparations are universally relevant and serve to further an agriculture that balances nature and culture for the benefit of humanity and the Earth.

11. João Volkmann, Capão Alto das Criúvas Farm, Brazil

Dr Ambra Sedlmayr, Dr Maja Kolar

Introduction

João Volkmann is a rice grower and currently one of the main producers of biodynamic preparations in Brazil. He was recommended by the Biodynamic Association of Brazil for this study. João Volkmann has been making preparations for about twenty years and has been running preparation-making courses on his farm since 2001.

João Volkmann is based on the farm Capão Alto das Criúvas in Rio Grande do Sul, the southern-most state of Brazil. This region is classified geographically as pampas and has a very humid sub-tropical climate.

Dr Ambra Sedlmayr and Dr Maja Kolar visited the farm of the Volkmann family from May 9–13, 2015. On the first day a visit to the rice fields took place, followed by the in-depth interview. In the afternoon the preparation storage was visited and the preparation practice interview was started. On the second day this interview was continued. During the remaining days further visits to the farm took place and details concerning the preparation work were clarified with João's daughter, Gabriela Volkmann. No practical work with the preparations took place during the visit.

Farm portrait

João Volkmann's father bought the 560 ha (1,384 ac) Capão Alto das Criúvas farm near Camaquã in 1954. The landscape of this large farm is gently undulating, with rice fields, artificial lakes and pampas vegetation in the wide valleys, and 250 ha (618 ac) of native forest and pasture land in the hills. The farm buildings and rice processing workshop form the heart of the farm.

The altitude of the farm is between 30–80 m (100–263 ft) above sea level. The weather in the region is humid and very unpredictable; there can be droughts or floods at any time of year. The average annual rainfall is 1,200 mm (47 in). The native forest helps to moderate the farm's microclimate.

The soil in the area is of granitic origin. It is acidic, has an excess of aluminium, limited amounts of phosphorous and medium levels of potassium. The fields in the valley were drained for rice production some eighty to ninety years ago. They have an alluvial soil that is heavy, dark, fertile and rich in clay, and may be compared to the soils at the bottom of a lake.

João's father was acquainted with anthroposophy and as soon as he bought it, started work on conserving the landscape of the farm. Between 1974 and 1983, however, the farm was rented out to a conventional farmer and João Volkmann remembers that when he took over the farm it had been a 'chemical desert'.

When he started farming, João thought he should establish a crop rotation, alternating rice with soya beans and maize. After five or six years, however, he realised that the soil was only suitable for growing rice – the soil lost its fertility when it was drained for other crops and the frequent natural flooding affected the production of maize and soya. João therefore established a system of alternating rice production during the summer months with cattle grazing in winter.

A rice field at Capão Alto das Criúvas is inspected before the harvest.

There is a herd of 120 cattle on the farm and another herd of 60 water buffaloes with their offspring (nearly another 60 animals). The water buffaloes are well suited to the region, since they deliver their calves between January and April when there is a lot of grass available. They are also well adapted to the humid environment.

João currently manages 200 ha (495 ac) of rice fields, 70 ha (173 ac) of which are rented from neighbouring farmers. Rice is therefore the main source of farm income. It is sold to a large number of small health food stores all over Brazil and to other places in South America. Cattle and water buffalo are sold for meat. In the future, some wood will be available for sale too.

After the harvest and during winter, cattle graze on the rice fields.

João Volkmann and his wife are supported in their work by their children. Two of them are actively involved with the development of the farm, and the others help with certain areas of work. In addition to the family, the farm provides work for fifteen people.

First steps in preparation practice

João was born in 1959 and grew up in Porto Alegre in an anthroposophical family. João used to spend his weekends and holidays on the farm Capão Alto das Criúvas. From early on he knew that he wanted to become a farmer. When he was fourteen he had the opportunity to spend a month at the biodynamic Demetria farm near Botucatu and experience biodynamic agriculture and work with the preparations. João had known about the preparations as a child. He remembers how, when he first became involved in making and applying them, 'It felt very right. Like, this is how it has to be. As a teenager I always had a strong wish to make the preparations.'

João had many opportunities for participating in lectures related to anthroposophy and biodynamic agriculture. His parents often hosted the speakers who would give lectures in Porto Alegre. João explained how the anthroposophical content always felt quite natural to him: 'From every person who came to give a lecture, I always learned a lot … There is something about anthroposophy, it seems to reveal a knowledge that we already have inside ourselves. It is nothing new, one comes to find agreement with it – yes, this is how nature works. And it seems that Steiner provides the keys for us to understand how things are.'

João Volkmann, preparation maker and farmer at Capão Alto das Criúvas.

João was trained as an agronomist. In 1983, when he took on the farm, his intention was to farm it according to biodynamic principles. This was a gradual process that took several years to achieve. He began by focusing on building up the basic farm infrastructure. Only when the de-husking of the rice could be carried out on the farm were direct sales possible, and only then did Demeter certification become relevant. The farm has been certified since 2001.

During the first three years of his work on the farm, João sometimes had to apply herbicides. This felt to him like 'a stab in the heart', but the conditioning of his agronomic training caused him to believe that it could not be done any other way. João started to make the 500 and the 501 preparations and this helped him to stop using herbicides. Through his trust in the preparations he took a leap of faith and stopped using agrochemicals.

In 1996, João travelled to Germany in order to visit biodynamic farms and, as a result, 'was inspired to make all the biodynamic preparations.' He met Christian von Wistinghausen and attended a preparation making course on his farm in the year 2000. In 2001, João invited Christian to Brazil to run a preparation-making course on his farm. It was an experience of great joy and satisfaction for him to be able to make preparations on his farm. The course with Christian was the first of the annual preparation-making courses during which all the preparations are made on his farm – all, that is, except Valerian, which does not grow in the region.

João said that it was not something someone told him that made him want to make the preparations, it was something that had been living within him for a long time. 'Very crazy,' he commented. João feels that, having established the biodynamic principles and gradually developed the farm organism, his work with the preparations could become more intensive.

How the work developed

João stated very clearly: 'I am of the line of Christian von Wistinghausen. I'm a pupil of his.' According to João, Christian von Wistinghausen was very demanding and strict regarding the way preparations should be made. João agrees with this approach and takes it 'very seriously'. He aims to continue making the preparations as Christian von Wistinghausen taught him. He does not believe himself wise enough to make significant changes and also feels satisfied with the results he has achieved so far in applying what he has learned.

While diligently applying what he has learned, João has adapted some practices for rice cultivation. He has developed techniques for simplifying some of the tasks. One of João's inventions has been to hang bags filled with compost treated with the biodynamic preparations in the irrigation channels that feed the rice fields. He does not have enough compost to spread over all the rice fields, nor does he think there is a lack of nutrients. João developed this practice in order to bring the effect of the compost preparations to the fields so that they can guide the decomposition processes that are active there. Since the 500 and the 501 preparations are carried by water, he thought that the effects of the compost preparations could be carried by water too. João is sure that this system is effective. As well as placing bags of compost in the channels that lead into the fields, João also gives frequent applications of Cow Pat Pit preparation (CPP) to the rice fields to introduce the effects of the compost preparations.

In terms of practical improvements, João has taken up the practice, learnt from a visiting student, of washing out the brain from the skull using water pressure from a hose. He also found it easier to break open the skull with an axe in order to take the preparation out, instead of laboriously scraping it out with wire. Other detailed adaptations are described in the section on preparation-making practice.

In João's experience, work with the biodynamic preparations is a never-ending journey of discovery, and he continually comes up with new ideas, insights and questions to pursue. João believes that there is a lot still to be discovered regarding the potential uses and applications of the preparations. He feels that there is great need for research in support of biodynamic farming. Farmers have many questions, but they have neither the know-how nor the means for doing the necessary research themselves. One question that could be researched, for example, is what other plant species could be used for preparation making in Brazil. There are a number of valerian varieties growing in Rio Grande do Sul and one of them is quite similar to *Valeriana officinalis*. João once made valerian juice from it, but was unsure about using it since he had no means of determining whether it worked the same way as *Valeriana officinalis*. João said that to address such questions it would be helpful to have a Goethean Science Research Institute.

João would like to see more places practising biodynamic agriculture and making the biodynamic preparations in future. He believes that what really matters is that people 'do it!', that they get into a rhythm with it and make the preparations every year at the right season. João wants to 'encourage more people in Rio Grande do Sul to use the preparations, and we are helping this to happen.' His focus is on working close to home, 'because here where we live, we can accompany the projects and people with more care.'

João Volkmann's understanding of the preparations
General understanding

João's experience with the biodynamic preparations leaves him in no doubt that the preparations are effective. He said: 'The preparations must be used, otherwise there is no way of knowing whether they work or not.' When converting to biodynamics and establishing a balanced farm organism, it is very important in his eyes to focus on the preparations.

For the context in which João works, the most important reason for using the preparations is to secure a dependable harvest of high quality rice, despite annual environmental and climatic variations. This attaining of resilience is not only an ecological feature, it has a deep effect on how João approaches his work. He finds calmness and tranquillity through using the preparations. He explained how he becomes insecure if he has not applied the preparations to a field, and how he is filled with trust that all will be well if he has been able to work with the preparations in ways that have been shown to be effective. João speaks of biodynamic farming as an 'agriculture of trust' in contrast to an 'agriculture with poisons' that he sees as being 'an agriculture of fear – you are always afraid of pests, afraid of everything, always, always…'

João observes that some people who are new to biodynamics want to start by making peppers of the various pests in order to control them. He believes that such a focus on peppers originates from the fear of pests and diseases that is the hallmark of conventional agriculture. João recommends first of all working with the preparations and bringing the farm organism into balance.

João explains how the preparations create resilience through their complementary effects. They can act in one way or another, depending on the particular environmental circumstances, to achieve greater harmony within the farm organism. He uses the example of humus whose complementary effects can make a sandy soil more compact and a compacted soil more open.

João described how the biodiversity of his farm has increased since he started working with the preparations. He said: 'It is something we can read directly, if we open our eyes. The preparations seem to bring about some reconciliation between agriculture and ecology.' There is also a beneficial effect on the relationship of people to the farm: 'There is a great love for this project among the people working here. I think this is part of the farm's "ego organisation". The new generation, my children wanting to stay here, is a reflection of this.'

João believes that the preparations can be used to help regenerate a damaged ecosystem, and sees an important role for them in the future in healing land that has been damaged by conventional agriculture and industry.

João is convinced that in accordance with the intentions of the Agriculture Course, food that is produced with the help of the biodynamic preparations can strengthen human willpower. He has seen new initiatives develop in places where his rice has been introduced. He said that he often receives 'love declarations for his rice' and hears stories from people who believe they have been cured of an ailment by eating his rice.

Preparations as information transmitters

According to João, the preparations are transmitters of information and the quantity used is of secondary importance. He used the example of traffic lights, where the colour red tells the cars to stop. In that situation one does not speak about 'quantities of red'. João said that the quality of the information is what matters not the quantity. The preparations need to be 'very well made' so that the right information can be given to nature. Making the preparations conscientiously and 'exactly in the way Steiner indicated' is therefore of paramount importance if the right kind of information is to be produced.

The information contained in the preparations is one reason why João feels that the recommendations on preparation usage should not be overly focused on quantities, but rather on their qualitative aspects. He also feels that the ideal of the self-sustaining farm organism should be considered when recommending how much to use. The current Demeter recommendation of applying 300 g (10½ oz) of the 500 per hectare (2½ ac) cannot be fulfilled using the amount of horns which João could produce on his farm (10–20 horns per year). João feels bound by this recommendation, however, and therefore obtains horns from a slaughterhouse to produce sufficient 500 for his farm.

Rather than being thought of as a form of abstract or numerical information, João feels that the preparations need to be understood artistically, like painting or music. He compares the various preparations to musical notes and stresses the importance of using them all, because 'you cannot play music with only one note.' Or again: 'Preparation work needs to be done with the heart. One needs to understand what one is doing. The moment you understand, you fall in love with it.'

Storage considerations

For João, it is important that the preparations are well insulated during storage so that their radiant power does not dissipate. In the moist climate of his farm, the preparations remain loose and have a low but constant moisture level. João learned from Christian von Wistinghausen that excess moisture can bring about undesirable processes or cause the premature activation of the compost preparations, which should only happen when they have been inserted in the moist compost heap. Anaerobic processes occurring during the period of maturation in the soil, or during storage, result in poorer quality preparations. They develop 'in the direction of silage' rather than undergoing 'the transformation process typical of the preparations.'

Working with the preparations in the tropics

It is clear to João that in his region and throughout the southern hemisphere the preparations need to be under the soil between Easter and Michaelmas – during the winter period. The cow manure produced on his farm from the ripe autumn grass in April has a nice structure, whereas the fresh spring growth in September produces a very watery, greenish manure. He feels that the Christian calendar is less relevant than the rhythms of nature; in fact, João believes that it would be better to adapt the dates of the Christian festivals to the seasons of the southern hemisphere.

João is open to the possibility of native plants being found for making the preparations in Brazil. He feels blessed in being able to grow most of the preparation plants in his region with the exception of valerian, which can still be grown at higher altitudes. João does not see any ecological problem with growing the exotic preparation plants in Brazil because they are only needed in small quantities, do not cause damage, and work 'like medicines for the earth.' For João any use of substitute plants would need to be supported by long-term research based on the Goethean approach.

Preparation making on the farm

João thinks that it is very important for biodynamic farms to get into the rhythm of making their own preparations each year and in accordance with the indications given by Steiner. He has found that farms on which the preparations are made are likely to have a stronger commitment to the biodynamic impulse, whereas farms that merely buy the preparations often revert to organic production methods. 'Where the preparations are made, there is a culture of understanding for the essence of biodynamics. And this gives greater consistency,' João explained.

Understanding the field spray preparations

João considers the 500 and the 501 preparations to be polar opposites and yet both are, in his view, 'preparations of the sun.' He explained that the sun is still active on the Earth at night and that the 500 has an effect similar to the sun at night: 'The sun then draws the roots of the plants down from the other side of the earth and during the day draws the aerial parts of the plants upwards, making them grow vertically.'

The importance of horn manure for humus formation

João believes that it is very important to understand the vital processes taking place in the soil. The soils in Brazil tend to be acidic, contain excessive amounts of aluminium and only a limited amount of phosphorous. By increasing the soil's humus content much of this problem can be solved. An increase in soil organic matter of 1% will increase the pH by half a point, thus a soil containing 2% organic matter can rise from an initial pH of 5 to a pH of 6. In a soil that is richer in humus, the aluminium binds itself to humic substances and is no longer available in the soil and hence no longer toxic to plants. In order to keep tropical soils fertile it is vitally important to increase and maintain humus in the soil at a high level. João feels that this is where the preparations have something very special to offer. He said: 'Humus is everything. It is very, very marvellous. And here the biodynamic preparations are hugely important, because they teach nature how to make humus'. Levels of organic matter in the soil of Capão Alto das Criúvas have increased due to the preparations. João ascribes this largely to the 500.

João described how the 500 has a relationship to the ego forces of the Earth: 'The grass that the cow eats grows vertically out of the earth and this is connected to the Earth's ego forces. It is as if the cow were eating this I of the Earth and concentrating it within herself during the 18–20 hours of her digestive process. And then I take this further and give these forces the opportunity to concentrate themselves further inside the horn over winter.' In João's opinion, it is this relationship with the Earth's ego forces that enables biodynamic food to increase the will forces of those who eat it. João believes that what happens inside the horn during winter, is a concentration of these forces. Microbiological processes are of minor importance in this. He explained: 'These are the little secrets of the biodynamic preparations. If we were to research the various bacteria and fungi contained in the preparations we could be researching for thousands of years. But it is not about the bacteria, it is about the forces … And the human being is given the task of guiding these forces.'

João believes that the forces held inside the 500 are concentrated to such an extent that in order to 'awaken them' and make them active in the processes of nature, they must be brought back into a living rhythm through the process of dynamisation. Because of the spiritual and chronobiological significance of the hour, this process of dynamisation needs to last for exactly one hour 'not 59 minutes nor 1 hour and 1 minute.'

In his understanding, the 500 works in the region of the plant root, at the interface between the living plant and the mineral soil – a transition layer made up of mucilage and micro-organisms. The plant interacts with this layer of living soil and this in turn mobilises the nutrients required by the plant during its stages of growth.

Horn silica preparation

In João's opinion, the 501 preparation is a very special substance that distinguishes biodynamic agriculture from all other forms of farming. João said that the physical substance of a plant is created to a great extent through its photosynthetic activity and only about 3% is the result of minerals being taken up from the soil. It is therefore extremely important to 'fertilise the aerial part of the plant' in order to support healthy plant growth.

In João's experience the 501 preparation gives an impulse for leaf formation. The 501 also increases the shine on the rice leaves. They are typically yellow-green in colour and this colour is enhanced. This contrasts with the bluish-green that occurs following nitrogen fertilisation. This colour difference is noticed not only by João but also by his workers and neighbours. Whereas nitrogen fertilisers keep the plants in a vegetative process, the 501 fosters the ripening process, something that is largely suppressed in conventional agriculture said João.

There is a school of thought that advocates not using the 501 in Brazil, since it might exacerbate the already very high light intensity. João disagrees with this point of view, since in his experience the preparations have harmonising effects. He could also imagine that the 501 dampens down the effects of excessive light. João has observed the indications given by Corinna von Wistinghausen that the 501 can be used to help plants cope better in times of drought. João tried this on his pastures and felt that it had a positive effect. It gave justification to the notion of there being a harmonising effect with horn silica.

Making sense of the compost preparations

To João Volkmann's understanding the compost preparations help to regulate the nutrient cycles within the farm organism. They also help to activate, retain and recycle the nutrients on the farm. A similar effect occurs in the compost heap, where nutrients are held together in the process of humus formation.

João relates the compost preparations to the planets and tries to understand them individually by pairing them according to the principle of 'complementary antagonisms', similar to what has been indicated by Jochen Bockemühl and Kari Järvinen in their book *Extraordinary Plant Qualities for Biodynamics*. He thus connects oak bark to the Moon, which stands in polarity to valerian which is connected to Saturn. Nettle (Mars) and yarrow (Venus) are opposites – nettle is male and 'speaks', whereas yarrow is female and 'listens'. Chamomile (Mercury) and dandelion (Jupiter) form another pair – Mercury dissolves and Jupiter gives shape.

João is convinced of the effectiveness of the compost preparations. The main effect that he has observed since he started placing bags of prepared compost in the irrigation channels, is a change of smell in the rice fields after they are flooded. There is no longer any smell of putrefaction, but a softer smell 'like in a stable'. There are also less pests in the rice fields since the compost preparations have been regularly used.

The social setting

Since 2001, two courses on the biodynamic preparations have been held on the farm each year. Around thirty to forty people attend each course. The positive experience of holding these courses has led to the development of other seminars and further educational activities, including hosting Brazilian and Latin American biodynamic conferences. The farm workers are also trained to use the preparations.

João Volkmann also supplies preparations to many other biodynamic initiatives, whose development he also supports more or less closely. Some state run agricultural research institutions have also purchased his preparations. In this way an active exchange of practices and ideas has been fostered by producing the biodynamic preparations on the farm.

Preparation practice

The preparations are made each year during the four day preparation-making course that takes place in March or April. The dates chosen are selected to suit potential participants and generally run from Thursday to Sunday. During the course, a cow is slaughtered and hung up on a tree. Anatomy is then taught by direct observation and the organs needed are extracted directly before use.

Cosmic rhythms and the use of the Maria Thun Calendar are very important for João, but he does not rigidly follow the Maria Thun Calendar on his farm since it is located in a region with very unpredictable weather. 'As a farmer I have to be pragmatic ... working according to the calendar is an ideal, but I must also be practical.' João therefore prefers to apply the preparations when the appropriate stage of plant growth has been reached, rather than waiting for the 'most correct date' to do it.

Field spray preparations
Obtaining and handling horns

Since the 10–20 horns that João can obtain from his own cattle per year are not enough to produce the quantities of preparations needed for his field and to supply other farmers as well, João needs to get additional horns from elsewhere. João goes in person to a slaughterhouse to select and saw off the horns from carcasses. Since in Rio Grande do Sul most of the cattle are kept on pasture, João feels that the horns are of good quality. Traditional craftsmen also use horns, and there is a certain amount of competition and also understanding for the value of horns. João exchanges the horns for rice. The horns can be used up to three times, after this they become too thin and weak.

Horn manure preparation (500)

The manure is collected directly from the pasture and from animals that do not receive any additional feed. The manure is stuffed into the horns in March, around Easter time, when the days start to shorten and the forces of nature begin to withdraw.

Most of the course participants, including João, use their hands to fill the horns. Afterwards lemons are provided for washing hands, since soap would fix the smell of the manure. The horns are tapped firmly on the table to make sure they are well filled. Between 760–770 horns are filled with manure each year.

Having had problems with water logging in the past, João now chooses a sloping site to bury the horns and takes care to keep it well drained. The floor of the pit where the horns are buried is also sloping. A drainage ditch is built around the pit to divert excess water. This ditch also prevents tree roots from growing into the preparations.

The horns are buried at a depth that corresponds to the height of the horns and then an additional 20–30 cm (8–12 in) of soil is piled on top of the horns. This depth is chosen because the soil is alive and aerated in the surface layer. Deeper down there would only be mineral soil. The horns are placed in the pit in the same position that they had on the cows' heads – with the tips pointing upwards. This is also to prevent water filling the horns. Some 15–20 horns are placed in the pit at a time and carefully covered with soil.

The 500 is taken out of the soil around Michaelmas. João checks the quality of the preparations in the first samples that he takes out of the soil. He checks their quality by observing their colour and smell. According to him, the 500 'should have a slightly sweet, "cow" smell.' If the preparation is still 'too green' he leaves the whole batch in the soil for a bit longer.

The preparation is removed from the horns by gently hitting them upside down against the wall of a bucket. If something sticks inside, it can be taken out the next day using the same procedure. The 500 is then made into balls, corresponding to the amount of preparation to be used for one hectare (2½ ac) of land. João takes care to ensure that these balls do not dry out in the store. To this end, water can be added to the peat surrounding the wooden box containing the 500.

Horn manure preparation is applied while the rice seeds are germinating. Three days after sowing, the water is drained off the paddy fields and the 500 is applied. The entire field is sprayed using a knapsack sprayer that produces a small water jet that extends up to 5 m (16 ft) in each direction, with droplets falling everywhere from the sprayer to that point.

When João first became Demeter certified, he was applying 60–80 g (2–3 oz) of the 500 per hectare (2½ ac). Now, in order to comply with the Demeter regulations, he applies 250–300 g (9–10½ oz). Even with a good balance of animals, pastures and fields, João could never produce the amount of horns needed for this quantity of the 500 within the farm organism, which is an issue that bothers João.

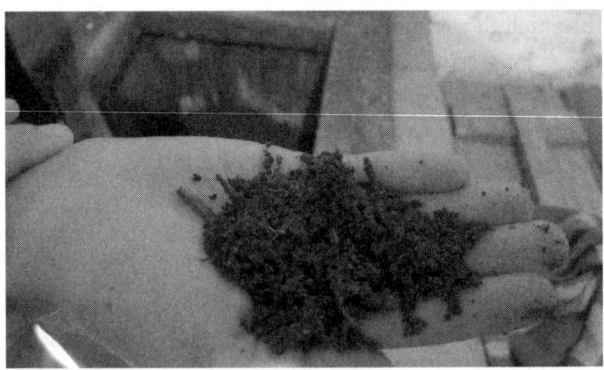

The 500 at Capão Alto das Criúvas is moist and loose.

Horn silica preparation (501)

The horns for making the 501 are only used once. This follows a recommendation given by Christian von Wistinghausen. João made nine silica horns in 2014/15.

Amethysts are used and these are collected from an area north of Porto

Alegre where they can be found by the roadside after roadwork or as waste from factories that process crystals. João has been advised to use local feldspar for making the preparation, but he is unsure about its effectiveness and therefore prefers to continue using amethyst.

The amethysts are first crushed in a metallic cylinder using an iron pestle. The iron is later removed using a magnet covered with toilet paper (for easy removal of the iron from the magnet). Finally, the crushed material is ground up very finely between two pieces of granite.

The pulverised amethyst is made into a 'little soup' by adding water. It is then put into the horns, which are placed in a box filled with sand. After two days the excess water will have evaporated and the horn openings can be closed with a layer of clay. The horns are buried at a place on the farm on top of a hill where they can receive sunlight from morning till evening. The horns are buried with their tips upwards, as they were on the animal. The horns stay in the soil from late September to March.

When it is dug out, the 501 is quite moist. The preparation is stored in transparent glass jars on an outdoor shelf on a wall of the preparation store.

For spraying, 5–6 g (1–1½ tsp) of the 501 are used per hectare (2½ ac). The preparation is applied on the pasture and at least twice on the rice fields each year. 501 is applied early in the morning. It is sprayed on the pastures and on the rice plants as their leaves emerge from the water when they are some 10–15 days old. Since the paddy fields are flooded when the 501 is applied, the spray can only be applied from the margins of the paddy. Care is therefore taken to ensure that the wind carries the spray across the fields.

Amethyst crystals are used for the production of the 501.

Before harvest, the rice fields are again sprayed with the 501. This time the application is done in the afternoon to further the ripening process. João remembers Christian von Wistinghausen giving this indication. His own explanation is that the day can be compared to the four seasons, with the morning corresponding to spring, noon to summer and the afternoon to autumn. If he wants to further vegetative growth, the 501 should be sprayed in the morning (spring), if he wants to further ripening, it is sprayed in the afternoon (autumn).

João believes that the 501 gives the young rice plants 'an impulse towards leaf formation' and the leaves get a stronger and shinier coloration. When the rice leaves lose their vigorous colour, the farm workers know that the 501 needs to be applied again. The end of season spray carried out in the afternoon, results in 'the straw looking healthier and the rice husks getting a stronger, much prettier yellow colour.' If the 501 is not applied at this stage to advance ripening, there is a chance that fungi will develop on the rice husks and turn them dark.

Stirring and applying the spray preparations

The preparations are always stirred by hand. João uses a wooden barrel that holds up to 180 l (47 ga) of water and a bamboo stick to stir. The preparations are stirred for exactly one hour. João always starts stirring around the edge of the barrel and moves gradually inwards as the vortex develops. He always stirs in such a way that the bamboo stick moves towards the heart – when stirring towards the left-hand side he uses his right arm, when making a vortex towards the right-hand side he uses his left arm. João experiences stirring on his own as a meditative process. He said: 'Looking at the vortex is enchanting, one enters almost into a trance, I like it very much.'

Stirring is often a shared experience with people taking turns in stirring. It is a time when many new ideas can develop. When there are other people present João makes sure that stirring is not accompanied by any ritual (for example, such as singing, praying or movement), and is carried out in a very calm and straightforward way. He thinks that making a ritual of it would be a distraction from the actual work with the preparations. His concern is that rituals could cause confusion in the future; the rituals could be passed down and become more important to people than the work of stirring.

João times his application of the preparations according to the growth cycle of the plants. Even though 200 ha (495 ha) of rice are cultivated, the spray preparations are only applied on a small area of 5–10 ha (12–25 ac) each time. 501 may be applied on up to 20 ha (50 ac) in one go. João believes that the 500 and the 501 are 'almost like one preparation', and using one of them on a given field requires that the other one is also used, otherwise imbalances could ensue.

João Volkmann demonstrates the application of the 500 with the knapsack sprayer.

Compost preparations
Yarrow preparation (502)

The stag bladders are bought in from preparation makers in Germany or Switzerland, since deer are strictly protected in Brazil. João tried to use bladders from Argentina, but they were from young animals living in captivity and the bladders were not strong enough.

Yarrow is grown on the farm in the ornamental gardens and along the tracks bordering the rice processing unit. Yarrow plants flower very intensively from December and keep on producing flowers for a long time. There are just enough flowers at the end of November to fill bladders directly with fresh flowers. The stag's bladder is first moistened with yarrow tea. Then an opening is cut into the top to fill it. Afterwards this opening is sealed with sisal twine.

The stuffed bladders are then hung up outside the house. They stay there until March or April when they are buried during the preparation-making course. The bladders are not moistened before being buried because the soil is very moist. When they are buried the bladders are protected on four sides by clay tiles and with wood in the base and on top of the pit.

'One needs to be a bit of an archaeologist' when taking the bladders out, said João. The bladder itself has normally disappeared, but the yarrow remains as a ball and needs taking out carefully. João has always been satisfied with the quality of the yarrow preparation and has not experienced particular problems with it.

Chamomile preparation (503)

Fresh intestines from a home-killed cow are used. The intestines are emptied out but not washed.

Chamomile is grown on the farm. It has to be dried and stored so as to have the flowers available in March when the intestines are filled. João's nephew, who is currently growing herbs on one part of the farm for tea production, provides the dried chamomile needed.

The chamomile flowers are first moistened with chamomile tea and are then stuffed into the intestine using a funnel specially made by João's oldest son. The intestine is placed on the funnel, the end is tied together and filling begins.

The intestines containing chamomile are buried in the middle of the farm, close to a house. They are protected on all sides with clay tiles. The chamomile sausages stay in the soil throughout the winter and are taken out in spring. The preparation is easily removed from the soil in the form of whole sausages, since the intestine stays intact. The chamomile preparation is sometimes too moist when it is taken out. It is stored in a clay pot, surrounded by peat.

Nettle preparation (504)

Nettles are grown on the farm. They grow well but need to be cared for so that the native vegetation does not suppress them. The nettle flowers most intensely in December. The preparation is therefore made at that time, rather than during the preparation-making course.

A large quantity of nettles is cut and left to dry in the refectory for three days. They need to be turned from time to time to prevent fermentation taking place. In João's experience, the nettles need to dry first so as to become a kind of hay and prevent them getting too dark or even forming silage in the soil. This is necessary due to the high level of soil moisture.

For burying, a pit is dug in the garden by the house. The hole is lined with wood along the sides and with a thick layer of peat on the bottom. It is then thoroughly filled with the dried nettles. On top of the nettles a layer of peat is added, followed by a layer of soil. The nettles stay in the soil from December until the December of the following year.

When it is dug out, most of the peat can be removed before the nettle is found. It is normal for some of the peat to get mixed in with the nettle preparation. The nettle is completely transformed within a year; in some years there is almost no sign of any stalks left in the final preparation. This depends on the climate and the intensity of the biological processes in the soil.

Oak bark preparation (505)

The skull for making the oak bark preparation comes from the cow that is sacrificed each year for making the preparations.

João's grandfather worked in the public forestry services of Rio Grande do Sul and grew the oak trees that have been planted in many a town square. Oaks grow very fast in the Brazilian climate. João can now collect branches when the trees are pruned and use their bark. He uses a manual grain mill for grinding the oak bark. The bark is ground so finely that the bark fibres are loosened and form a cotton-like foundation to an otherwise powdery brown substance. João produces ground oak bark for preparation makers all over Brazil.

João believes that the meninges (the brain skin) plays a very important role and so fresh skulls are used. João explained the role of the meninges by saying: 'The meninges is the skin that reflects the activity of the brain and hence makes "reflection" possible. The brain depends on this reflective activity and on its retention by the meninges.' João associates these qualities of reflection with the Moon forces.

The brain is removed from the brain cavity using pressurised water from a hose. The oak bark is filled into the meninges using a funnel and by stuffing the bark into it with a wooden stick. This year, João started using a piece of oak wood, carved to the right shape, to close the opening in the skull.

The skull filled with oak bark is placed in a very damp and swampy area where an artificial lake is used to channel water into the rice fields. Wood is put on top of the site where the skulls are buried and thorny plants are used to protect them from wild animals. In spring, the skulls are taken out of the swamp and opened with an axe to take out the oak bark preparation. The preparation is always very moist. It is placed directly into an unglazed clay pot and stored.

To João's understanding, the oak bark preparation relates to the calcium processes. Whereas the chamomile preparation draws in calcium via the sulphur process, the oak bark preparation draws calcium forces in via silica. These two preparations are particularly important for managing fungal attacks on the farm.

Dandelion preparation (506)

João has previously used sections of the greater omentum but has more recently taken to using the mesentery tissue. This envelopes the intestines and is almost transparent. João thinks that both work, but would appreciate some research to definitively show which part is best.

Dandelions rarely flower on the farm. But in the nearby towns it grows wild and whenever a member of the Volkmann family goes to town in the right season, they bring back some dandelion flowers. There is also an exchange arrangement with a farm to the north of Porto Alegre whereby dandelion flowers are carefully harvested for João in return for some rice.

When making the dandelion packages the dried flowers are first moistened with dandelion tea. They are then sewn up into packages using sisal twine. They are buried in a hole lined with clay tiles.

When they are dug out, the packages are always fully intact. The fat in the mesentery seems to hold it all together and prevent the total decomposition of the dandelion flowers. When the more fatty greater omentum is used, the dandelion flowers are usually slightly less transformed.

Valerian preparation (507)

Valerian preparation is bought in, since *Valeriana officinalis* grows but does not flower in the region. There are eight varieties of valerian growing wild in the area and one of them is similar to *Valeriana officinalis*. However, since João has no means of testing whether it works in the same way as the classical valerian, he has not adopted the use of the local variety.

Applying the compost preparations

Compost is mostly composed of rice husks. The rice husks are first placed in the area where the animals sleep for one or two months, so that they are manured and trampled on. After one or two months the husks are piled up into a heap with the help of the tractor. The preparations are then inserted. To do so, little clay balls are made to hold each of the preparations. Each clay ball contains as much preparation as fits between three fingers. The reason for making clay balls is that the rice husks are very loose and the preparations would not stay in the middle of the pile by themselves. The clay balls help the preparations to stay where João wants them to be. Since clay is a relatively conductive material and does not insulate, the preparations can take effect.

A bamboo stick is used to make holes in the compost heap and insert the individual clay balls. Four preparations are inserted in the sides of the heap, and the nettle is placed in the middle. Since the compost heap can be rather long, several sets of compost preparations are used. João always inserts a clay ball containing nettle preparation in between two sets of preparations 'to make the connection.' João does this in accordance with the advice of Christian von Wistinghausen, who explained that the nettle preparation 'awakens' the other preparations. The valerian preparation is stirred for 15 mins and sprayed as a mist on top of the heap. It takes about a year until the compost can be used. During this time the heap is turned three times, with a new set of compost preparations being inserted after each turning.

To bring the effects of the compost preparations to the rice fields, bags filled with compost are suspended in the irrigation channels leading to the rice fields. The compost is also applied directly on poor soils using a tractor and the shaking disks normally used for spreading lime. In this way two tons (1,815 kg) of compost can be spread on one hectare (2½ ac) of land.

Burying and storing practice
Burying practice
The preparations are buried at least 80 m (260 ft) away from each other which, given the large size of the farm, João thinks to be 'relatively close by.' The soil is similar in all the places where the preparations are buried. Care is taken to find a site where they can be buried in living soil. With the exception of the skulls, the preparations are all buried in pits lined with wood or tiles. The sites chosen correspond as closely as possible to the indications given in the Agriculture Course.

Storing the preparations
There is a room used for storing the preparations within one of the farm buildings. CPP and 500 are each stored in a wooden box surrounded by peat and placed within a compartment made of adobe. The lids of the boxes can be held open using a chain attached to the wall. The compost preparations are stored in unglazed clay pots in a wooden box, surrounded by peat. João is convinced that the preparations lose forces if they are not protected and well insulated with peat. He believes that well-insulated preparations can keep for several years and only gradually

lose their effectiveness. João advocates getting into a rhythm and making the preparations every year and so generally keeps them for at most two years.

Derived preparations and teas
Cow Pat Pit preparation (CPP)

The use of CPP has become very important for the rice growing cycle at Capão Alto das Criúvas. For making CPP, fresh manure is collected from the pasture. This is mixed with basalt dust and ground eggshells. These ingredients are mixed together and water is gradually added. The water is bound up with the material and a well-aerated substance is produced that has a consistency of wall plaster. After an hour of stirring, the material is placed in a wood-encased hole in the garden. The hole is first half filled with the mixture and then the compost preparations are added. The rest of the CPP material is put on top and valerian, which has been stirred for 15 mins, is sprayed over it.

A ladle full of CPP is used for stirring in 60 l (16 ga) of water. Some extra valerian preparation is added to foster the activity of phosphorous that is deficient on the farm. This mixture is stirred for 15 mins. Three drums are filled with the dynamised CPP preparation and put on the tractors that are used to work on the fields. Each drum has a tap on the bottom. This is left slightly open so that the CPP solution can keep dropping on to the soil while the tractor is working. CPP is always used when harrowing, but not when ploughing.

CPP is applied some four to five times a year in the rice fields. This is because there is a lot of organic matter whose decomposition and humification needs to be guided.

Horsetail tea

João does not use horsetail tea in the paddies, since he grows rice varieties that are very resistant to fungi. Horsetail tea is sometimes used in the gardens. For this purpose horsetail is boiled in water for an hour. The decoction is then diluted ⅛ in water and sprayed with a knapsack sprayer whenever there is a problem with fungal diseases on a vegetable crop. The application of horsetail tea for three days in a row in the evenings, and repeated weekly, has cured tomato plants of botrytis. Preventive application on the soil is helpful if the fields are known to be prone to fungal attack.

Summary

Anthroposophy has been part of João's life since childhood. Rather than experiencing a decisive change in direction, the will to work with the preparations lay dormant in João for a long time and gradually began to manifest itself. Being able to learn from Christian von Wistinghausen gave João the courage and the impulse to start making all of the preparations. Most of the preparations are made during the annual preparation making course that takes place on the farm.

João aims to keep his preparation practice unchanged and in accordance with what he has learned from Christian von Wistinghausen. He works in a calm, conscientious and simple way when making and applying the preparations and avoids any 'flourish' (such as accompanying the stirring process with a ritual). João likes to be practical and pragmatic. He has adapted his preparation practices to make the tasks more efficient without losing accuracy (for example, opening the skull with an axe to take oak bark preparation out), to suit the local conditions of the farm, and to support the cultivation of rice. Particular innovations include hanging bags containing prepared compost in the irrigation channels and using CPP every time the tractor does any type of harrowing in the paddy fields. These practices have been developed so that the effects of the compost preparation can be brought to the paddy fields that do not receive compost.

The high levels of humidity in the area require João to take special measures to keep the environment where the preparations-to-be are buried from becoming anaerobic and to prevent undesired fermentation processes. The preparations are stored as soon as they come out of the soil. The insulation of the preparations during storage is considered to be of utmost importance in keeping them effective.

Being an agronomist, João has studied the life of the soil with great attention to detail, and learned about the importance of humus in the chemistry of tropical soils. He feels that the preparations are key to guiding the processes of decomposition and humus formation and that biodynamic agriculture is therefore the only approach that holds the secret of maintaining and building up soil fertility.

João has many questions with regards to the potential use of the preparations and the use of substitute plants and minerals. He believes that sound, Goetheanistic research is needed to answer the questions of farmers who are not able to conduct this research themselves.

João has a heartfelt connection to the preparations and to rice production. He makes a practical study of them and does so with great interest and care.

Rice production is the main focus of the farm and the preparations are used in its service. The preparations provide a foundation of trust for João. They give him confidence that the farm organism is balanced and able to cope with erratic circumstances while maintaining a steady and consistently high-quality yield of rice.

12. Devon Strong, Four Eagles Farm, USA

Anke van Leewen, Dr Maja Kolar

While this case study was being finalised the sad news of Devon Strong's passing in late November 2015 hit us. We hope that our portrayal of his work can make his impulse live on.

Introduction

Devon Strong is a biodynamic farmer and producer of preparations in northern California. In adapting preparation making to North American conditions he uses organs from his own herd of American bison (Bison bison, also known as buffalo). He also uses some local plants for making the preparations. His work is influenced and informed by his study, and by experience of Native American culture.

The biggest producer of preparations in the USA is the Josephine Porter Institute in Virginia. The preparations in the United States are mostly made by individuals without group participation. The biodynamic group in Oregon, of which Devon Strong has been a member for about twenty years, is an exception to this. This group has developed a special culture of mutual exchange regarding the making of preparations.

Devon Strong's Four Eagles Farm is located in Montague in North California, close to the border with Oregon. The landscape of the Mount Shasta valley, has been shaped through volcanic activity. The area has an upland desert climate.

Dr Maja Kolar and Anke van Leewen visited the farm on September 21 and 22, 2015. On September 21 there was a farm tour and an interview concerning preparation practices. The researchers participated in the making and burying of horn manure, chamomile, yarrow and dandelion preparations. The in-depth interview took place on September 22.

Farm portrait

In the early eighties, Devon Strong started farming in Ashland, Oregon, where he leased 16 ha (40 ac). There he developed a CSA garden and also kept sheep. He moved his Four Eagles Farm project twice, before buying some land in Montague in 2003. This is a wide-open pasture land. He found water and installed his own well. At the moment there are no permanent structures on the farm, apart from a few hay barns.

Buffalo herd grazing at Four Eagles Farm.

The farm covers 89 ha (220 acres) of flat land and is situated 914 m (3,000 ft) above sea level. The rainfall in this area is about 480 mm (19 in) per year, concentrated in the winter and spring (between November and May). Summers are very dry and hot with an average maximum temperature of 33°C (92°F) in July (usclimatedate.com). In winter, snow and frost is possible. The soil is heavy clay and of volcanic origin. Most of the land is used as pasture apart from 10 ha (25 acres) of irrigated hayfield, and a small market garden. Devon Strong keeps about twenty buffaloes; this includes cows, the herd bull and their offspring. There is also a flock of triple purpose sheep with about twenty ewes, three horses and chickens. The butchering of sheep and buffalo is done by Devon himself on the farm. The meat is sold by direct marketing. Devon uses Native American ceremonies to deal with and communicate to his animals, especially the buffalo.

Market gardening is carried out on 1 ha (2½ ac) of the land and seeds are produced for home use. Garden produce has been certified as Demeter LOCAL since 2015. This means that it has to be sold within a 200 mile radius of the farm. This type of certification is based on local groups, using a peer-to-peer verification system (demeter-usa.org).

Devon Strong is not a full-time farmer. He also provides farm services like sheep shearing and horse shoeing and runs workshops on crafts such as butchering and wool processing. There is currently one other worker on his farm, who is mainly responsible for the vegetable garden.

First steps in preparation practice

Devon Strong was born in Nevada in 1956. His father was a rancher and he was raised on cattle ranches. As a child he was introduced to all kinds of farm work, like cattle breeding and using farm equipment. Out in the country he often saw signs of the way indigenous people used to live and wondered how they were able to survive in such a forbidding desert environment. Most of the Native Americans he knew, did not know a lot about it, and the formal description of their way of life, did not explain how it was possible to live in this harsh desert.

After high school, Devon decided that he wanted to work with animals, so he started a degree in animal science at the California Polytechnic University in San Luis Obispo. He soon found that the technical and artificial way he was being taught to deal with animals was not what he was looking for. So he left the college, having gained a horse-shoeing certificate, and worked for several years as a cattle rancher.

In his twenties he met a native elder in Ashland and was introduced to the Native American culture of the Lakota tribe. He discovered the spiritual connection these people had with the land. Since then he has been practising and studying the 'native ways'. He is accepted as a relative by a Native American family in South Dakota and leads sweat lodge ceremonies for various communities.

Devon began farming on leased land in the early eighties. During that time, while running a CSA vegetable garden, he was searching for ways to combine organic gardening with Native American ceremonies. This led to the practice of holding a sweat lodge ceremony during the full moon period to bring female energies into his garden.

Once, when Kathryn Casternovia from the Oregon biodynamic group visited his farm, he told her that he was trying to bring native ways and organic gardening together. She suggested that he should consider working with the 'European spiritual approach to farming', called biodynamics. When Devon Strong heard that biodynamic agriculture works with spirits called gnomes, undines, sylphs and salamanders, he realised there was a connection to the native ceremonies, where stone, water, air and fire are represented. For him this was 'just another language for the same thing.'

He was soon convinced that biodynamic agriculture was what he was searching for. He bought the Agricultural Course and read it the first time by himself.

In 1991, he started to work with the Oregon biodynamic group, where he deepened his understanding in exchanges with the other farmers and also had his first experiences making the preparations.

The preparations give him the possibility of connecting with the farm in a way other than simply doing the daily chores. His favourite aspect of the preparation work is the work with animals and their organs.

For Devon, biodynamic agriculture is also offers the possibility of compensating for the mistakes made by former generations – a kind of intergenerational atonement.

How the work developed

Devon Strong's reflections on the preparations were, from the beginning, shaped by his Native American perspectives and his strong connection to animals. When he first visited the Oregon biodynamic group during their preparation-making session, he was very impressed and quickly realised that he wanted to make the preparations himself. He saw a correlation between the native way of making offerings to the spirits and the use of biodynamic preparations. For that reason, it was obvious that he should make the preparations himself. He explained: 'I don't want somebody else making my prayer ties for me'. (A prayer tie is a small bundle filled with tobacco that is used as an offering to the spirits in Lakota culture.)

He joined the Oregon biodynamic group, which meets four times a year in order to make the preparations in spring and autumn. There is one meeting for reading and storytelling in the winter and a meeting to evaluate the preparations in the summer. In their summer meetings, the group takes six to eight preparations from different locations and observes, compares and discusses them in great detail. A blind test is made, where all the farmers make personal notes about the preparations and then afterwards they share their impressions with the group.

The first time Devon made the 500 himself was in 1993. As he had no cow horns available, he made a trial using sheep and goat horns. Although the substance seemed to be very similar to other 500s, the Oregon biodynamic group members dismissed it. They told him that goats and sheep horns are not suitable, because they have an outward expression of energy and do not hold it inside in the way cows do. They also encouraged him to enhance his spiritual perception of these qualities. By holding it in his hand, Devon could feel that his preparations had less energy than the 500 made in cow's horns. This was also a moment when he felt reassured that biodynamic agriculture was the right way of farming for him. He felt attracted by the spiritual understanding underlying it. He said to himself: 'Oh my god, this is a whole different level of agriculture.' He appreciates that other people are working with scientific approaches to understand the preparations in order to complement the spiritual perception, which is his main interest.

It was the event in 1993 at the Oregon biodynamic group that also inspired Devon Strong to study the archetype of the cow. He found that the buffalo, which shares the same ancestry as the cow, has exactly the same qualities. He explained: 'When I first heard the description of the archetype of the cow, I thought it was a description of a buffalo … the hair and horns – that is the buffalo.' In 2000, he began to make the 500 and the 501 with buffalo horns. Some of the Oregon group members were very impressed because the preparations seemed to be full of energy, whereas others had their doubts as to whether wild animals are suitable for making the preparations. Devon kept working with buffalo horns and organs, and this particular approach was gradually accepted by the biodynamic community in the USA. He also attracted the interest of the International Working Group on Biodynamic Preparations and was invited to the Agricultural Conference at the Goetheanum in 2010 and 2015.

Devon Strong feels that American soil has many variations in temperature and temperament. The buffaloes seem to be more adapted to this environment than cows. He says: 'They (the buffaloes) have an inherent value for the land, they carry something very powerful for the land. You do not have to be a native person to know that.'

Devon is currently producing nearly all of the preparations on his farm and he sells them in small amounts to between forty and fifty gardeners all over the country. His vision is to spread his work with buffalo and biodynamic agriculture all over North America, including his use of native ceremonies to deal with and connect to the buffalo. In his opinion, the biodynamic preparations can help to re-enliven and restore the landscape of North America through their connection with the physical and spiritual qualities of the buffalo. About buffalo in general he thinks: 'Honouring the traditions that have connected buffalo and people for thousands of years, gives all Americans a sense of connection to this land we love so much.'

Devon Strong's understanding of the preparations

For Devon Strong, using the preparations is an invitation to the elemental beings to help and work on the farm. Like in the native tradition, Devon pursues a way of life that acknowledges a daily relationship with the spiritual aspects of all things. His intention is to 'bring the quality of life to the level that allows the spirits to interact with it all the time.' Physical gestures, such as working with the preparations, can reinforce this relationship.

Devon compares preparation work to the Native American tradition of making prayer ties. These are small bundles of tobacco that call the elemental beings to work with or for humans. In the native tradition, tobacco is a powerful plant that is able to hold the essence of a prayer. Different colours on the sheath of a prayer tie attract different kinds of spirits. Devon wonders if different coloured sheaths for the preparations would create different qualities. For Devon, the preparations are like prayer ties, but much more specific. They attract a certain kind of spirit. He thinks that the particular combination of preparation materials and the use of the preparations sets an intention, whereas the prayer ties receive their meaning through human prayers. Devon believes that the preparations help in the transformation of spirit into matter and matter into spirit. In his opinion the use of local materials makes it easier for the spirits to connect with the farm. He explains: 'The whole aspect of having the preparations attuned to the area where you use them, I think, makes it easier for the spiritual applications to apply to the specific farm. It is a more general offering to the spiritual world, when you offer chamomile from somewhere else or from overseas to the spirits in your area. It is still an effective offering, but the depth of quality in its relationship to the farm is enhanced when more local materials are used.' He therefore prefers to use local plants, like the local chamomile and the local subspecies of the European nettle.

Following native traditions, Devon avoids working with the preparations during the middle of the day, since he believes that the spirits do not interact with humans during this time, as they also do not during the middle of the night.

Making sense of the field spray preparations

For Devon Strong, the 500 and the 501, translated into native language, represent the mother and father influences. The 501 is the father influence, connected to the sun, and the 500 is the mother influence, connected to the Earth.

When Devon Strong first moved to his present place and started working in the desert climate, he did not use the 501. He did not want the strong sun influences to be increased further, and tried to bring more of the 'composting earth senses' to the parched land. Then he realised that this was a mistake and that he needs the horn silica to educate the sun influences: 'They were just wild and had too much influence. So I started using it and I got what was missing. I was afraid to use it because I did not want to bring more of it. But this brought control to it.' That is why he thinks it is important to use both of these polar preparations. In order to help control the sun influences, he started to bury the 501 for a whole year so that it can experience the whole cycle of the sun, thereby learning more about it.

In his perception the 500 is 'a concentrated energy that enriches the whole environment.' It also has a strong grounding effect on him. The 501 seems to be the opposite and is connected to light and the cosmos. He describes his experience with crushing crystals: 'When you first break them, there is a huge release of energy that is dizzying and takes your mind out to the stars. It is extreme enlightenment and it just lifts you up from the earth. There is a huge polarity there, both in the making of the preparations and the stirring.'

According to Devon, the release of energy that occurs when the crystals are crushed, seems to free up elemental beings. He compares it with making homoeopathic medicine. He says: 'I identify it with the idea of incarnated beings who come into the form of an apple and how, when you bite into it, you are releasing them. When you break the crystals you are releasing the elemental beings. They are joyful and they are now more available to help. And making the 501 in a finer and finer way is almost … like the homoeopathic remedies.'

Making sense of the Dandelion preparation

Devon Strong thinks that the dandelion has a very fine and internally directed energy. The function of the animal sheath is to protect this tender quality from outside processes in the same way as it is protecting the digestive process inside the animal.

The Oregon group members do not like to use very fatty organs, as they think that the fat insulates the dandelion from outside influences. Their focus is on the membrane that envelopes the intestines and they believe that this is the part needed to obtain a high-quality preparation. But Devon thinks that the mesentery can have too little fat. He considers: 'Maybe the fat actually has a really protective quality to it ... I feel like it allows an instreaming of qualities, that we do need to go into the dandelion. I think we are dealing with far finer influences than with yarrow or valerian, which are externally extremely powerful.'

In his opinion, the organs of old cows have the best quality because they do not have as much fat as those of young animals. Old cows are seldom available to him, though, because in his breeding system he sells off cows instead of changing the bull.

The effects of the preparations

To Devon the effects of the preparations are not immediately observable, but have a rather more subtle quality. He explains that he sees their effects through the reaction of people, when, for example, they come to his place: 'It's when other people come to the farm and say: "Wow what a wonderful feeling there is here." And then I look at all the junk I have here and all the things lying around and I think: How can anybody think that this has anything wonderful about it?' The garden produce also seems to have a special quality and often draws a positive response from customers.

Devon Strong has a special relationship to the nettle plant and he uses it in many different ways – in the form of nettle preparation, as a fermented tea, and he also uses a lot of nettles in the barrel preparation. For him, the nettle brings the quality of feeling and sensibility to the compost and also has a great formative influence.

The social setting

Devon Strong mostly works alone on his farm, but invites people to join preparation work as often as possible. He participates in the meetings of the Oregon biodynamic group, and for several years has joined in discussions with the national working group on biodynamic preparations in the USA and travelled all over the country to attend their meetings. Sometimes, experts on the preparations like Dennis Klocek or Matias Baker join his preparation making days.

Devon offers various farm services and these occupy a large part of his working time. Through this work he meets many farmers and takes these opportunities to spread awareness and understanding about biodynamic agriculture. He has a special connection to the Native American societies and once or twice a year visits reservations in South Dakota, where he also makes preparations with a Waldorf school.

Preparation practice

The eight classical preparations recommended by Steiner are produced as well as a number of additional ones, such as CPP, horsetail tea, fermented nettle and blood manure compost, all of which are produced and applied on the farm. So far, the preparations have been used in the garden and on the hay field and on some pastures that are close to the farm. Devon plans to expand the use of preparations to the whole farm. To this end he has just bought (in 2015) a field sprayer.

Devon would like to use the Maria Thun Calendar for preparation work, but from practical experience he finds that it is not possible to organise his work in that way.

Field spray preparations
Obtaining and handling horns

To make the field spray preparations, Devon uses horns from the cows of his own herd of buffalo and some buffalo horns obtained from a butcher. Buffalo horns differ from cow horns. Young and old, female and male horns are easily distinguished. Devon explained the difference: 'The young ones are thin and shallow and old horns are thick and very, very full of horn material. Male ones have a large opening and females have a narrow opening.' It means that old horns do not have much space left to fill and the horns are very solid and heavy in comparison to cow horns.

Before using them for making the preparations, Devon cleans the buffalo horns and, if necessary, also washes them. He uses some buffalo hooves as well, although he thinks that the quality of preparation made using horns is better. Horns and hooves are both used for several years. Devon Strong keeps horns used for making the 500 separate from those for horn silica.

Devon explained how in the western states of America it is not possible to distinguish between male and female cattle horns once they have been removed from the animal. It is usually possible to identify female horns by the stress rings that appear on the horns due to having had a calf. But he observed that in the western states of the USA, both male and female cattle develop stress rings on their horns during the harsh winters. Even butchers have confirmed to him that they cannot see a difference between bull and cow horns. This is something he believes should receive more attention from preparation-making groups, especially if the origin of the horns is unknown.

Differences between old (left) and young (right) horns.

Horn manure preparation (500)

Devon uses manure collected the same day for making the 500. It comes from the female animals on the pasture. The manure has a paste-like consistency and seems highly concentrated. He fills the horns and also some hooves by hand or sometimes using a small wooden stick.

He buries the 500 in the centre of the farmyard. When he moved to the farm in 2003, burying horn manure was the first thing he did and he still uses the same spot. He buries the horns in a horizontal position. When there is manure left over from filling the horns, it is put on top of the horns in the hole. 500 is applied at least twice a year in spring and in autumn when the soil is moist.

Horn silica preparation (501)

Devon Strong explained that the Oregon biodynamic group uses crystals from Germany to make 501. He prefers to use the white quartz that can be found in the region. For crushing the quartz, he uses a big stone mortar and a hammer. He finishes the milling process between two glass panes in order to get a very fine powder. He moistens the powder with water and makes a paste to fill the horn. He does not seal it with clay, because he wants the influences of the sun to come in.

Devon buries the filled horn on a small hill where it will be exposed to the influences of the sun. The pit is around 30 cm (1 ft) deep. He buries the horn in summer and leaves it in the soil for an entire year, until the following summer. Usually he makes one horn with silica each year.

Stirring and applying the spray preparations

Devon stirs and applies the field spray preparations by hand. He applies the 500 in the afternoon, just before dark so that the dew comes and 'ingests it'. The 500 is used at least twice a year, in the spring and in the autumn when Devon adds 180 ml (6 fl oz) to a 20 l (5 ga) bucket, which he then applies to his 1 ha (2½ ac) garden.

The 501 is sprayed only once a year in summer time at sunrise, but Devon thinks it would be better to use it with the same frequency as the horn manure preparation. Until now, he had no equipment for spraying so he used to sprinkle the field spray preparations by hand.

Compost preparations

Yarrow preparation (502)

Yarrow blossoms are collected on wild areas in the spring. Devon always does this work in the morning. The fresh flowers are put in a paper bag and dried on top of a refrigerator to make use of its heat.

For Devon it is difficult to receive stag bladders regularly. He sometimes gets a stag's bladder from a hunter. At other times he might take a bladder from a road kill animal. The local deer are black-tailed deer or mule deer. Black-tailed deer have a small bladder, with a diameter of about 10 cm (4 in). Devon explained that in the USA all kinds of deer bladders are used for making the yarrow preparation, even elk bladders are sometimes used. Devon makes sure that the bladder comes from a stag. He is also wondering whether it is possible to use a goat bladder and whether it would be effective on occasions when it is not possible to find a deer bladder.

Devon usually fills two or three bladders each year to produce yarrow preparation. He stores the bladders in the freezer. Before filling them he defrosts and blows them up. Devon moistens the yarrow flowers in hot water and stuffs them carefully into the bladder. It does not matter to him whether there are a few yarrow stems inside the bladder.

The yarrow-filled bladders are hung up on a tree where they remain over the summer. During the visit for this case study, which took place in autumn, Devon buried the bladder directly after filling it. Since he had not found a bladder at the beginning of the summer, he left making the preparation until autumn.

Devon buries the bladders inside a plastic growing pot with some soil around them. This helps him to find the preparation again, something he has found difficult in the past.

Chamomile preparation (503)

Devon uses the local chamomile, *Matricaria matricarioides*, also called dog fennel. He thinks that it has the same quality as *Matricaria chamomilla*, especially the qualities necessary for biodynamic purposes. Devon collects the chamomile blossoms in May, and always in the morning. He takes all the blossoms that are flowering without selecting a specific stage of flowering. The flower heads are dried in a paper bag on top of the refrigerator.

Devon prefers to use fresh buffalo intestine. He sometimes also uses a sheep intestine or takes some that have been frozen. Instead of cleaning the intestine with water, he only removes the content because he wants the bacterial layer to stay intact.

The dried flowers are moistened with hot water and carefully stuffed into the intestines with the help of a funnel to make sausages.

Nettle preparation (504)

For making the nettle preparation Devon uses a sub-species of *Urtica dioica* that grows in the area, *Urtica dioica subsp. gracilis*.

He collects nettles in autumn just after the equinox. Since he often harvests nettles for other purposes as well, there are still plants growing that have not developed seed. The tops of these plants, including stems and leaves are cut.

The nettles are left standing in water to keep them fresh before they are buried. Devon had previously separated the nettles from the soil using a window screen. For the last four years he has used a buffalo pericardium as a sheath, because many people associate the nettle with the heart. His perception is that the quality of the preparation is enhanced by this practice.

He buries the nettles for one year. According to Devon, some people in America leave the nettles in the soil for two years.

Pericardium taken from a sheep.

Oak bark preparation (505)

Devon Strong explained that there are two different types of oak, white and black, growing in the area around his farm. For making the preparation, he uses the white oak (*Quercus alba*).

Devon uses a horseshoeing rasp to remove the bark from the tree. He removes the crumbly outside part of the bark and uses the inner bark. He collects the bark immediately before filling the skull.

Since he has been raising buffalo, Devon always uses a female buffalo skull for making the oak bark preparation. In his opinion, it is ideal to use a fresh skull from an older animal, one that is at least two years old. He finds that the skull from younger animals can fall apart more easily. When he uses a skull of an old animal, he can use it several times. He fills the skull with the bark using a spoon and closes it with a wooden plug as it is hard to make a bone fit. Devon has also used sheep skulls and has been satisfied with the results.

Dandelion preparation (506)

Dandelion flowers are collected from the meadows around the farm. The flowers are harvested on a day in March, before noon. Devon takes flowers that are open and are at an early stage of development. He dries the dandelion flowers in the sun or in a closed bag in the shade.

The greater omentum from a buffalo is used for making the preparation. Dried flowers are moistened with hot water and put into the greater omentum. One big package is assembled by folding the greater omentum and tying it up with a cord. The preparation is then buried.

Devon Strong with a piece of greater omentum.

Valerian preparation (507)

Devon cannot grow valerian on Four Eagles Farm as it is too dry and hot there. He usually helps to make the valerian preparation with Don Tipping on Seven Seeds, in southern Oregon. They collect two or three big bowls full of flowers and squeeze them out with a high pressure steel compactor. They only use the corollas to make pure juice. The valerian juice obtained is stored in dark bottles. It lasts for two or three years, at which point a new batch of valerian preparation is made.

Applying the compost preparations

The compost heap is arranged in a round shape. The compost preparations are arranged following the numbers (502–507) in a circle close to the centre of the heap. Approximately 2 ml (½ tsp) of each preparation is added to a compost that is not bigger than 10 m³ (13 yd³). Around twelve drops of valerian juice are added to half a litre (½ quart) of water and stirred for 15 mins. Some of it is put into the hole and the rest is sprayed over the surface of the compost pile.

Burying and storing practice

Burying practice

The burying place for each of the preparations is carefully chosen. The compost preparations, except the oak bark preparation, are all buried in the garden where the soil is rich and treated with the preparations. The compost preparations are buried to a depth of 30 cm (1 ft) in a row of separate pits – one for each preparation. Devon marks the spots with some wood or a bone.

The skulls filled with oak bark are buried by the overflow of the water tank. Sometimes it happens that the tips of the horns that are left on the skull show up because of water erosion.

Devon buries most of the compost preparations around the autumn equinox. If possible, he waits for some rain, but in 2015 there was a serious drought in California and he had to irrigate the places where he had buried the preparations so that they had a good start. The preparations are dug out shortly after the spring equinox.

Storing the preparations

The preparations are stored at Four Eagles Farm in a wooden box and placed outdoors in the shade, near to where the horns are buried. The preparations are stored individually in glazed clay pots that are surrounded by peat. The wooden cover of the box also contains a thin (2–3 cm, 1 in) layer of coconut fibre. Devon takes care that the preparations do not dry out and moistens them with water from his well every six weeks, sometimes more frequently than that. Valerian juice is stored in dark bottles in the same box. The 501 is stored in a glass jar in the kitchen in a place exposed to the sun.

Derived preparations and other applications

Cow Pat Pit preparation (CPP)

For producing a CPP-like preparation derivative, Devon uses fresh manure and an equal amount of nettles and stirs the mixture with a shovel in a rhythmic spiral for an hour. He does not add eggshells or basalt flour. Half of the mixed material is put into a bottomless barrel and then one set of the compost preparations (502–507) is added in small holes. Then he fills up the rest of the barrel and applies the compost preparations once more. Stirred valerian preparation is applied on top. The filled barrel is covered with a stone and is left to mature over the winter, from fall until spring.

Devon uses this barrel compost to boost the effect of the 500 and applies it on the same places and with the same frequency as the 500.

Horsetail tea

Horsetail grows in southern Oregon along creek sides and Devon goes to collect it there. Horsetail is used as fermented tea as a preventive as well as a treatment for mildew in the garden. The tea is applied using a watering can at a concentration of 235 ml (8 fl oz) to 20 l (5 ga) of water.

Nettle tea

Devon Strong makes a fermented nettle tea. For this he fills a 20 l (5 ga) bucket with nettles and fills it up with water. The liquid nettle manure is used to reduce the stress of transplanting seedlings and also as a stimulator of plant growth. He uses a concentration of 235 ml (8 fl oz) of fermented nettle tea in 20 l (5 ga) of water.

Blood manure compost

The blood manure compost is a mixture of buffalo blood, manure and a small amount of water added to prevent the mixture drying out. This is covered for the composting process. When the fermentation has finished it has a black colour. Devon thinks that this is a highly concentrated and powerful fertiliser. He uses it directly in small quantities as a fertiliser or as a starter in the greenhouse.

Summary

When Devon Strong first heard about biodynamic agriculture and the preparations, he had already been studying Native American spirituality for several years. Native American spirituality has a great influence on his perception and on his way of thinking. Devon seems to pay the same attention to the spiritual world as to the material world. His understanding is that Native American culture and biodynamic agriculture fit together because they work with the same spirits. In his view, the use of preparations is an offering to the spirits or elemental beings and he compares it to the Native American tradition of making prayer ties. Devon relies predominantly on his spiritual perceptions and does not search so much for scientific explanations.

Devon always had a strong connection to the land and one of the key aspects of his preparation practice is the adaptation of preparation work to local conditions. He thinks, that using local materials for making preparations makes it easier for the spirits to understand the message and connect with the farm. For Devon, the most important aspect of making preparation practice fit in with the spiritual quality of the environment of the American continent, is to use buffalo horns and organs. He has the vision of spreading the work with buffaloes to biodynamic farms all over the country. Almost all the materials that are used for preparation making are produced on the farm or collected in the region. Some of the plants are relatives of the original European preparation plants.

His favourite part of preparation work is with the animals. As he is skilled in butchering, he is very familiar with all the organs. A unique aspect of his preparation work is that he started using an animal sheath, the pericardium, for producing nettle preparation. His focus on working with the preparations seems to be more in relation to the animal organs than the plants.

He has also adapted his practice to the desert climate. He buries the 501 for a whole year since he thinks that it will then contribute greater control over excessive sun influences.

Although Devon Strong introduced many innovations concerning the ingredients used in making the preparations and in the use of other additives, such as liquid manures and barrel compost, he does not seem to question the value of the classical preparations. He trusts in their effectiveness. Devon is a doer, a 'man of action' and it is more important to him 'to get it done', than doing it in accordance with the aim of achieving perfection. The preparation work and even spraying is done simply by hand with very little equipment.

13. Chris Hull, Hohepa Homes Community, New Zealand

Dr Ambra Sedlmayr, Johanna Schönfelder

Introduction

The biodynamic movement in New Zealand is still quite small, with only thirty Demeter-certified farms. Nevertheless, the movement is active on a national level and also engaged internationally, through its participation in Demeter International, for example, and by helping to promote biodynamics in India and other Asian countries. Chris Hull is the main producer of biodynamic preparations in New Zealand. He produces preparations for the Biodynamic Association, which then sells them throughout the country.

Chris Hull is based at the Camphill-inspired Hohepa Homes community in Hawkes Bay, on the south coast of North Island. The community was founded in 1957 by Sir Lewis Harris and Marjorie Allan. Lewis Harris was a wealthy farmer who had a daughter with special needs. This led him to support Marjorie Allan's initiative to start an anthroposophically inspired home for people with special needs by offering land for the establishment of Hohepa Homes. Five sites are currently managed by Hohepa community in various parts of New Zealand. At Hawkes Bay, Hohepa manages two connected sites that provide care for special needs residents: Clive and Poraiti, each with some 20 ha (50 ac) of farmland and gardens. Chris Hull manages the Poraiti farm together with his colleagues Andy Black and Jenny Lyons.

Poraiti is in a warm, temperate region with a thorn steppe biome, 40 m (130 ft) above sea level. The climate is mild, with annual temperatures ranging from 9°C to 19°C. January and February are the warmest months and July is the coldest month. There is no dry season. The average annual rainfall is 809 mm (32 in). The soil at Poraiti is made up of loamy sand and volcanic ash.

The researchers Dr Ambra Sedlmayr and Johanna Schönfelder visited

Chris Hull between March 26 and 28, 2015. They spent the first day getting to know Hohepa community by participating in a harvest festival and meeting biodynamic students from the nearby Taruna college. On the second day, Chris Hull gave a guided farm tour that was followed by the preparation practice interview, which took place in the preparation storage shed. An in-depth interview was conducted in the afternoon. No work with the preparations was carried out during the visit.

Farm portrait

The farm at Poraiti comprises 20 ha (50 ac) of land, with pastures, apricot and citrus orchards, greenhouses and gardens. It has been managed using biodynamic methods for 35 years. Previously it was owned by Sir Lewis Harris. When he died, the land was split into several plots and put on the market for sale. Chris already had a relationship to this land since he had been using it for his horses and spraying it with the 500 'to cure it'. He felt that it was important to secure this land for Hohepa community and together with colleagues he started a fund raising campaign. Through their combined efforts and enthusiasm they were able to purchase the land that makes up most of the current Poraiti farm.

There is a herd of shorthorn cattle made up of seventeen cows with their calves and some steers raised for meat. The cattle are actually de-horned, which is a source of ongoing controversy in the community. In New Zealand, the haulers and slaughterhouses often refuse to take horned cattle. It is legal to slaughter animals at home, under the so called home-kill system. The meat of the home-killed animals however cannot be sold. There is also uncertainty about how to carry out the de-horning. Is it better to dis-bud a calf at birth or let the cow live with its horns and then put it through the cruel procedure of de-horning it before sending it to the slaughterhouse? There is also a widespread perception that horned cattle are dangerous. The farm has three main production sectors: cattle, vegetables and fruit. Vegetables are grown on the fields and in greenhouses. They are mostly used within the Hohepa community, although some are sold. There used to be commercial seed production too, but it had to be stopped for a while.

The number of people working on the farm varies. The work schedule is arranged in such a way that there are always two workers and two residents on the farm every day of the week. There are usually also two WWOOFers (volunteers from Willing Workers on Organic Farms).

Chris Hull, preparation maker for the Biodynamic Association of New Zealand.

First steps in preparation practice

Chris Hull was born in 1951. He grew up in the suburbs of London, in the UK. In his late teens he travelled widely and then lived for a while on a remote island in Scotland. This awakened his interest in 'being involved with the land' and he dreamed of buying some land in Scotland or Wales and living a crofter's life. He then spent some time working in a children's home near London, doing gardening work with the children. During this time he came across the Emerson college in nearby Sussex, where he spent one inspirational day. This was when he first met Anthroposophy.

At the age of 24, Chris decided to follow his dream and find some work, perhaps in Australia, to earn the money needed to buy land back in the UK. During his travels Chris became very interested in the technologies and lifestyles he encountered. He was fascinated by simple and functional artisan work, with the possibilities of using what nature provided and with creating things for oneself. Chris eventually arrived in Australia where he found work of various kinds including construction and gardening work, and a short spell on a tobacco farm. He spent some time spent living in a very remote place in the Queensland rainforest. Chris then travelled to New Zealand and ended up living in the Northland region, where he stayed on someone else's land and experienced the way of life he had been seeking: 'Just living in nature and being involved with the land and developing it and making whatever you need.'

Chris developed a garden, used seaweed and fish to make his own fertiliser, and had a horse as his means of transport. Chris did this for some years and then decided he wanted to stay in New Zealand. To do so he needed a work permit. Two people he met advised him to go and work for Hohepa. After a first visit, Chris was offered a job and, when his work permit was granted in 1978, he left Northland and went south to Hohepa.

To begin with, Chris worked in a house community looking after adults with special needs and doing the gardening. This changed after less than a year with the arrival of two adolescents with Down's syndrome who needed supervision. These youngsters had to work on Clive farm. This is where Peter Proctor, a biodynamic pioneer, was working. Chris came to work with him and in this way was introduced to practical biodynamic agriculture. Chris vaguely remembered being interested in spirituality and biodynamics when he was in Queensland and Northland, but had not been sufficiently interested to take it further. Here at Hohepa he joined study groups, attended the annual anthroposophical conference at Hohepa as well as national biodynamic conferences, and immersed himself in reading biodynamic books. The anthroposophical images resonated with him and the acceptance he received from the people around encouraged him to explore things further.

After spending a year working with the two adolescent boys, Chris was asked if he could develop a gardening workshop on the Poraiti site for three other adults with learning difficulties. Chris agreed and, in doing so, was determined to use biodynamic methods to grow good quality food for the residents of Hohepa Homes and, at the same time, provide meaningful farming and gardening work for them. This was the beginning of the Poraiti farm, which Chris manages together with his colleagues to this day. Chris had to draw on his own skills and resources to build up the farm since there were no tools, infrastructure or capital to invest in it. He was left quite free to develop the farm with his colleagues in the way he believed right while balancing the needs of the community.

How the work developed

Chris Hull first got involved with preparation making when he worked with Peter Proctor on Clive farm in the early 1980s. What fascinated him when he first heard about the biodynamic preparations was that they 'took agriculture into a different dimension', beyond the material, everyday practice. It made him realise that there are many unknown and unseen things at work behind what one can normally perceive and that these are of the greatest significance. This opened up a whole new dimension of life for him. Chris recounted his initial enthusiasm when he said: 'You are working in such a different way than you normally work, picking these things at the right time and getting the animal sheaths and doing that ... It was quite special really ... And to actually see how they (the preparations) physically changed in the ground too. That was also something of significance for me.'

In his spare time, Chris went to help Peter Proctor make the preparations at Clive farm together with Clifford Hamer, who was at the time responsible for making the 501 preparation. This was in line with Chris' usual way of learning – watching people do it and then trying it himself. It was by this means that Chris had acquired many of his rural craft and self-sufficiency skills. This was how he now learned about the preparations: 'Nobody asked me, I was just interested and then I always follow these things up. Because there are so many things I had done in my life that I had not been trained in, like shoeing horses – you watch someone and then you do it. Leather work, I used to make a lot of shoes for my kids and lots of things. One has the idea and just does it. Just do it. Make mistakes ... It was the same with the preps. I was interested, you find out from the people doing it, and that's how I started.' Preparation workshops, led mostly by Peter Proctor, were additional opportunities for practical and theoretical learning.

Chris then took on the growing of the preparation plants as his responsibility and worked hand in hand with whoever was making the preparations at Clive farm. This was mostly Peter Proctor. Preparation making continued on a relatively small scale until Bert de Liefde, who was based in Wellington, announced that he was going to stop producing the preparations for the Biodynamic Association. The Association at that time was small, but interest in biodynamics was growing. It was clear to Chris that someone had to continue this work and he decided to take it on together with Peter. The preparations were important to Chris with regards to his work at Poraiti farm. He felt it important for other people to be supplied with the preparations and be supported in what they were doing too. It was a pragmatic decision. 'It was a job that needed doing,' he explained, and he took it on. The first income from the selling of preparation 500 to the Biodynamic Association went into the building of much-needed sheds for the two farms of Hohepa. It meant that the workers could finally have a place to go when it was raining and where they could store their tools. As Peter Proctor became more involved with lecturing and helping biodynamics to develop in India, Chris Hull took on full responsibility for making preparations for the New Zealand Biodynamic Association. Production was transferred to the Poraiti farm around 1990.

Preparation-making techniques have been refined and developed over the years and Chris has developed various tools for making his work easier. He created machines for grinding crystals, introduced the use of clay pipes to protect the preparations, and adapted a sausage machine for filling manure into horns. The preparations have always been made in the 'classical way' and kept in store in a reasonably moist state. No significant changes have been introduced to preparation practice, apart from the technical adaptations. This made Chris wonder whether he had been too conservative in his approach of just following the methods he had been taught without developing anything new. He explained that he had been satisfied with the way preparation making had been carried out when he took it on, and never felt the need to really question or change things, apart from his technical innovations. He reflected: 'I'm quite happy with the way I am doing it and maybe I would need some, not proof, but some evidence that another approach is better, I don't know. I know sometimes that when people bring out new methods it's because it's just easier. It's like instant coffee. I suppose I just feel very comfortable and feel that the way we are doing it in the traditional way is OK, you know? So maybe a question could be, why aren't I being progressive?'

Chris has always worked hard running the farm's commercial horticulture enterprise as well as taking care of residents. He also set up a seed production business to support other biodynamic farms. He has had a busy family life too, raising six children of his own. For Chris 'the preps were just another thing I was doing', so he didn't have a lot of time to reflect on them. It was simply a job that needed doing, amongst many other tasks. He has fewer duties now and is glad that he can spend more time working with the preparations, passing some of his know-how on to students and people in the workshops he is invited to teach, and have more time to reflect upon his work and experiences.

Chris is happy with his work making preparations for the Biodynamic Association for the time being, even though he is retiring soon and there is no successor yet. This does not concern Chris, because he knows that there are enough people who know how to make the preparations. He feels that when he stops, other possibilities could open up, such as the development of regional preparation-making groups, which have so far been difficult to establish.

Chris hopes that in future more people will take up the work and use preparations on their farms and gardens. For Chris, it is a question of how to motivate people since he does not perceive a great deal of interest in spiritual practices.

Chris Hull's understanding of the preparations

Chris never had any exceptional experiences through working with the preparations and is not able to trace any specific effects back to them. He explained that working on a farm where the preparations have been in use for over thirty years makes it difficult to perceive any objective changes after the application of the preparations, since no comparison is possible for him. For him it is more an inner feeling and an acceptance 'that these things are working.' This feeling came when he started reading and gaining an understanding 'of other beings and activities that are happening around us ... that we physically don't experience or see, but nonetheless feel are present,' and which Chris 'knew the significance and the importance of.' Chris acknowledged in his words that spiritual beings are working behind the apparent reality – it is an inner knowledge and trust that requires no further explanations.

In general terms, Chris' understanding is that the preparations enliven the soil and thereby enhance the vitality of the plants, and that this in turn results in better food for humans and animals. With regards to the 501, Chris believes that it generally strengthens plants and prevents disease. Visitors often experience a special atmosphere on the farm and comment on the high quality of the produce, both of which can be traced back to the biodynamic practices.

Chris believes that stirring the preparations by hand contributes human intention to the process and makes the preparations more effective. So although Chris has developed a number of tools and techniques to lighten the work of preparation making, he still feels the need to put a lot of dedicated work into them, which he sees as positive. Even though he acknowledges that cosmic pipes might be effective for radiating out the effect of the preparations without physically applying them to the land, they don't resonate with Chris: 'For me, working with the preparations is something that we tangibly do.' It is this connection, this being actively engaged with them, that has most meaning for him. This is also why his trial of using flow forms to stir the 500 or the 501 instead of stirring them by hand in a barrel, was soon abandoned.

As regards the preparation-making ingredients and preparation quality, Chris remembered how, in his work with Peter, touching and smelling were really important. Chris learned to make and store the preparations in a fairly moist state. He believes that this is important since water is the carrier of life and in this slightly moist state the preparations seem more alive to him. Chris has in this way taken on the approach learned from Peter Proctor, but he has also listened to his own feelings, having in past occasions made the decision to throw away preparations that had dried out.

Chris is not very open towards some theories and practices promoted within the biodynamic movement, by people who seem to get some personal gain out of promoting these. Chris feels strongly that the preparations should be accessible to all and that any control over them is inappropriate. Chris and his colleagues from the Biodynamic Association have also been asking themselves whether it is at all ethical to make preparations for sale, and they have tried to encourage regional biodynamic preparation-making groups to develop. These, however, have not yet become well established.

The social setting

At Hohepa community, work with the preparations is mostly undertaken with the help of the residents. Chris spoke about the joy and satisfaction of the residents when they come home with a basket full of chamomile flowers, knowing that they have contributed to something important. The residents also enjoy stirring the preparations, with everybody taking turns in stirring. Chris explained: 'When we stir 500 or make 500, everybody is involved and everybody has a part to play. And that's really great too.' Through the involvement of keen people, working with the preparations becomes a festive, joyful event.

Since the mid 1990s Chris has started making the preparations at Poraiti farm together with students from the biodynamic training course at Taruna college. There are two intakes of students per year and students have three eight day residentials at Poraiti farm. Adapting work with the preparations to suit the timetable of the course has meant placing less emphasis on choosing days with favourable constellations. Working with the students has turned preparation making into more of a learning experience, with questions and exchange.

Due in part to New Zealand's culture and the big distances between biodynamic farms, there is currently very little active exchange among biodynamic farmers and preparation makers. Sharing the work and experiences and offering mutual support is something that Chris would like to see happen more in the future.

Chris Hull with Tessa, one of the residents.

Preparation practice

Work with the preparations is adapted to the seasons of the southern hemisphere. In New Zealand, preparations are made in autumn (March–April) and taken out of the soil in spring (September–October).

Field spray preparations
Obtaining and handling horns

The horns are obtained from various sources. They mostly come from a home-kill butcher known to Chris who collects and passes on the horns. Other members of the Biodynamic Association also donate horns when they have some available and don't need to use them themselves. The horns are used for four to five years or until they physically deteriorate to the point that they can no longer be used. Since it is difficult to obtain horns in New Zealand, bull horns are not rejected for preparation making. Some trials have shown that the 500, when made in bull horns buried together with cow horns, is of acceptable quality, similar in consistency and stability to the normal horn manure preparation.

Horn manure preparation (500)

The manure is collected from the paddock at Clive farm. It is sieved through an old bed frame into a bath tub in order to get the grass out. The manure is quite fluid. Horns are filled using a sausage machine adapted by Chris for this purpose. The filled horns are left for a while to dry before being placed upright in the soil with their opening facing downwards to prevent them filling with water. The horns are placed side by side, without touching each other. In the past Chris has made some 200 kg (440 lb) of preparation 500 per year, in 2014 he produced 80 kg (175 lb).*

The days for stirring and applying the preparation are chosen according to the Maria Thun Calendar. Chris chooses a day when the moon is descending and is in an earth sign because then 'the forces are more concentrated in the soil.' Due to the work rhythm of the residents, who leave the farm at 3.30 pm, stirring needs to start rather early at 1.00 pm. A big ball of about 1 kg (2 lb) of the 500 is used for a 150 l (39 ga) wooden barrel. The stirring is done with the residents and sometimes other helpers. Six barrels are needed to spray the whole farm, this means about 285 g (10 oz) per hectare (2½ ac) are applied.

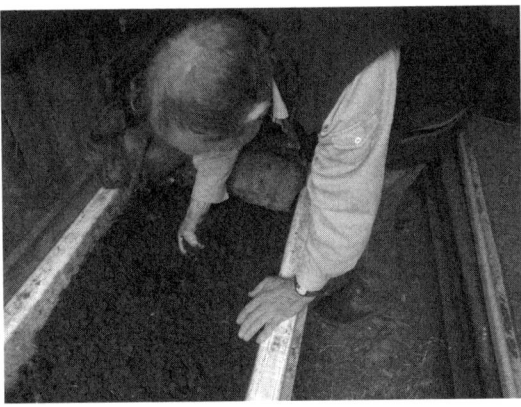

Chris Hull shows the 500 in storage.

---

* For certification purposes in New Zealand preparation biographies have to be produced. These contain the information of how much preparation in weight has been produced, not the number of horns or other organs that have been buried.

Horn manure is applied twice a year, in spring and autumn, in the greenhouse it is applied more often. Chris feels that it is best to apply the 500 when the land is moist; when it is too dry the soil does not seem able to absorb the preparation in the same way.

Horn silica preparation (501)

For making the horn silica preparation, clear quartz crystals from South America are used. At some point, Chris had been advised to use quartz sand instead to cut short the long process of grinding crystals, but since he wanted to provide the Biodynamic Association with high quality 501, he felt unsure whether quartz sand would be good enough.

For grinding the crystals, Chris follows a three step process. The crystals are first crushed in a metal pestle and mortar. The resulting pieces of crystal are sieved and the finer portions then go into the grinding machine developed by Chris. This machine works like a slow mill, with the crystal pieces being rubbed between two disks of glass. After taking the ground quartz out of the machine a further step of hand grinding is needed to get the powder fine enough. A magnet is used to filter out any metal particles that have mixed in with the quartz during the grinding process.

Chris selects the best horns for making the 501. The quartz powder is mixed with water to form a paste 'slightly runnier than toothpaste'. The horns are then left for about a week to allow the paste inside to solidify and for the water to be poured away. 501 is stored in glass jars in the sunlight. In 2014, Chris produced 10 kg (22 lb) of 501.

The 501 is sprayed in early summer after preparation 500 has been applied. A day is chosen for spraying the pasture just before it is closed up for hay making. A day with the moon ascending in a leaf sign is chosen. For spraying the fields, a 150 l (39 ga) barrel of 501 is stirred for an hour. The water is taken from a bore hole on the farm and is warmed up using a wood-fired cylinder. The 501 is then applied with a tractor mounted sprayer.

The gardens and greenhouses receive two horn silica treatments per year. For this 1 g (½ tsp) per 12 l (3 ga) of water are stirred in a bucket. Depending on weather conditions and the crop to be sprayed, it is spread over various sized fields. A 12 l (3 ga) bucket of the 501 is enough for half a hectare (1¼ ac) of pasture land. Because of the many different growth stages of the crops in the garden and greenhouses, Chris often finds it difficult to find the right time to spray the 501, since he does not want small seedlings to receive the 501 impulse before they have taken root properly.

Stirring and applying the spray preparations

The preparations are always stirred by hand with the help of the residents, sometimes WWOOFers and students also join in. Normally three barrels are stirred at the same time and people take turns in stirring. Two rounds of stirring are needed to spray the whole farm with the 500 using the three available barrels each time. Large fields and gardens are sprayed by Chris using the tractor, whereas smaller areas and terraces are sprayed with buckets by helpers.

Compost preparations
Yarrow preparation (502)

The stag bladders come fresh from a commercial abattoir. They are inflated using an air compressor and hung up in the glasshouse to dry. Once dried the bladders are deflated and stored. These bladders are also sold to Indian biodynamic initiatives. Yarrow has been sourced from different places in the past, including wild harvesting on the farm itself and from plants grown in the Weleda gardens.

Growing yarrow on the farm has not been very successful. Chris currently harvests yarrow from the wild. He discovered an area with plenty of yarrow near a beach and a river, in a seemingly uncontaminated spot. The flowers are harvested in November or December with their stalks. For drying, the flowers are cut off near the top, to reduce the amount of stalk, but the aim is not to obtain individual flower heads. The dried flowers are stored in paper bags.

In September (spring) of the following year, the bladders are moistened with warm water to make them supple, and the yarrow flowers are moistened with tea made of fresh yarrow leaves. An opening is cut in the bladder and is then filled using a funnel. Once it has been well stuffed the opening is sown up again.

The yarrow bladders are hung up in trees around the place where the preparations are buried. In March, they are taken down and buried inside a clay pipe.

To take the preparation out, great care is needed because normally the bladder has disintegrated. The preparation should have 'turned' to be considered of good quality, this means the flowers must be broken down and there should be a slightly sweet smell. If the preparation is not ready because it has been too dry, it is buried again and the whole area where the preparations are buried is watered.

Chamomile preparation (503)

The intestines are obtained from a cow butchered on the farm. The intestines are cleaned with water and then cut into 20 cm (8 in) long pieces. These are bagged and put in the freezer until needed.

The chamomile is grown at Poraiti farm. It self-seeds and seedlings are transplanted into a bed. The residents pick the flowers using a metal comb. They need to visit the beds every few days to collect enough flowers over the season. Some 5–6 kg (11–13 lb) of flowers are needed every year. The dried flowers are stored in paper bags.

The intestines are stuffed in autumn, in March or April when the biodynamic students come. To stuff the intestines, first one end is tied, then the intestine is placed on a funnel and the dry chamomile flowers are stuffed in. The filled 'sausages' are then placed inside unglazed clay pipes with all the sausages piled together in a lump. These are then filled with soil, placed in the ground and covered with more soil.

When it is dug out, the preparation can be taken out directly from the clay pipes, the pipe itself can remain in the soil. This is possible because the intestines are usually still intact and because the sausages have all been bundled together in piles and only separated from one another with soil. On retrieving it from the ground, the chamomile preparation has a sweet smell. Chris observed that the chamomile preparation seems to attract moisture, even in the store.

Nettle preparation (504)

Nettle has been planted and it now grows half wild at the edge of the farm woodland. Nettles are cut early in January, just before flowering starts, using a strimmer or weed cutter. The whole plants are dried on racks. Once they are dry, the whole plants are stuffed into an unglazed clay pipe for storage. The filled pipes are then buried sometime around mid to end of March. When it is taken out of the soil, the preparation is sieved to remove any remaining stalks. The nettle preparation does not keep for more than two years.

Oak bark preparation (505)

About six fresh cow skulls from home-killed animals are used every year. Chris cuts off most of the meat and extracts the brain using water pressure from a hose.

The oak bark is harvested fresh from trees growing on the farm. To do this the bark is grated directly from the tree using a wood rasp – it is easier to rasp it this way than grinding it when the bark has been stored.

The preparation is made in March. The oak bark is transferred to the skull using a funnel. Afterwards, the opening is sealed by ramming a stone in.

Because there is no constant source of running water, the skulls are placed in two plastic barrels that are then covered with vegetation and water. Three skulls are placed into each barrel.

To remove the mature preparation, the skulls are split open with an axe so that the brain skin with the oak bark preparation inside it, can be taken out whole. It is initially very smelly and needs to dry a bit until the smell has gone, this can take up to two weeks. Only then is the preparation put in the store. The consistency of the preparation should be light and without any lumps.

Barrel in which the skulls filled with oak bark are placed over winter.

Dandelion preparation (506)

The mesentery of home-killed animals is used. The mesentery tends to be quite fatty. Mesenteries are hung up in the glasshouse to allow the fat to drip off before use. It is possible to store them until needed at room temperature in a plastic container for about four to five months.

Dandelion grows wild on the farm, but to ensure that sufficient quantities can be harvested two beds have been planted. These beds are used for two years or until the weeds start suppressing the dandelions. The dandelions are then transplanted into new beds. Dandelion flowers are harvested when they still have 'a closed centre', that is, closed flowers in the centre of the flower head. Harvesting is done in the morning, between 8.30 am and 9.30 am. The flowers are dried on racks in the shade. When dried, the flowers are stored in paper bags. They are sorted before using in order to select those suitable for making the preparation (and to discard the ones that have gone to seed).

The dandelion pouches are assembled in March. For several years this has been done with the help of the biodynamic students. The pieces of mesentery are made into square pouches 20–30 cm (8–12 in) in size and filled with the dry dandelion flowers.

The dandelion pouches are buried directly in a pit in the ground that is clearly marked with sticks so that the preparation can be found again. The balls come out fairly intact. It is important that the flowers have turned, meaning that they are not recognisable any more.

Valerian preparation (507)

Chris grows a patch of valerian plants in his private garden. In November or December, when in bloom, the flowers are harvested by pulling them off the flower heads by hand.

To produce the preparation, the flowers are placed in big glass jars that are then filled up with water. To prevent the flowers rising up in the water, they are pushed down with a stone. The jars are then left for four to five days in the sun until the water has taken on an amber colour. The liquid is then sieved into a stainless steel bucket and finally transferred, without the sediment that forms at the bottom of the bucket, into jars for storage. The valerian preparation is stored in the dark.

The valerian preparation is made every year. The quality of the preparation is judged by its smell. It keeps for four to five years, and it is important to prevent oxidation. To do this, Chris makes sure he stores the valerian in full bottles with very little air inside. He pours the valerian preparation into ever smaller bottles after using some of it, so that it is always stored in a bottle containing as little air as possible.

Applying the compost preparations

There is a small paddock built into the pasture where the cattle are fed. This is where manure is collected for building compost heaps. When the compost heap has been built, the preparations are inserted. Small balls of compost are formed and 4 g (1 tsp) of preparation are placed inside. One set of prepared balls is then inserted in a row along each side of the compost heap. 15 ml (3 tsp) of valerian preparation are stirred in 10 l (2½ ga) of water for 10 mins and then sprinkled on the compost. The compost is usually only prepared once, but occasionally it is inoculated again after turning.

Burying and storing practice
Burying practice
All the preparations (apart from oak bark) are buried in the same area of the farm, between the shed and the greenhouse near to some trees, in a free draining area. Preparations are buried preferably during the descending moon and in an earth sign. Yarrow, chamomile and nettle are placed in unglazed clay pipes surrounded with soil and then buried. Dandelion pouches are buried directly in the soil. Wooden poles mark the exact location where a preparation is buried.

Storing the preparations
Behind the Poraiti farm building, there is a little wooden shed for storing the preparations. Here, large quantities of the 500 are stored in an enamel bath tub surrounded by peat. Some jute bags and wooden planks are placed on top. A wooden box, lined with plastic to prevent its decomposition is filled with peat to surround the clay pots containing the compost preparations. During the summer months, Chris checks the moistness of the preparations each month. If they are getting too dry he sprinkles some water on top of the preparations and mixes it in.

Derived preparations and other applications
Cow Pat Pit preparation (CPP)
In New Zealand, many biodynamic farmers make their own CPP, hence Chris does not need to produce big quantities for sale. At Poraiti farm, CPP is currently being made by the biodynamic students. Half a bath tub of cow manure, four cups of ground organic eggshells and four cups of basalt rock dust are mixed together for 15 mins using shovels. This mixture is then put into wood-lined holes in the ground and left for approximately three months. CPP is used in combination with preparation 500. Chris puts about 1 kg (2 lb) of CPP into the barrel where the 500 is stirred, half way through the stirring process of the 500. Seedlings in trays are immersed in a CPP and seaweed solution.

Horsetail tea
Equisetum leaves are boiled up in water to produce a foliar and ground spray that is mainly used for tomato and cucumber plants. Horsetail is considered a noxious weed in New Zealand, hence its transportation and use are at the margins of legality.

Summary

Preparation making at Hohepa community bears the hallmark of Chris Hulls' pioneering 'do it yourself' spirit and the constant presence and help of the special needs residents.

Chris was fascinated by the preparations from the beginning because, for him, they took agricultural practice into a different realm – that of unseen forces and beings whose presence Chris could acknowledge. Being a doer, it was important for him that the biodynamic preparations have a practical and a spiritual significance.

Interest in this work led to him becoming an apprentice with Peter Proctor and Clifford Hamer. He learned by watching and helping.

Chris' relationship to the preparations has been marked on the one hand by his certainty about their significance, and on the other by his practical focus on improving the technical efficiency of the work.

Chris has mostly continued to work with the approach to preparation practice he has learned and does not feel the need to question or change the practices passed down to him. He feels comfortable and satisfied with the current approach.

Regarding preparation practice, it is notable that chamomile, dandelion and nettle are used as dry ingredients when the preparation is assembled. That the skulls containing the oak bark are left in standing water is also unusual and a practical approach for dealing with the absence of running water on the farm.

Chris' has always kept the needs of others in mind and has a pragmatic approach in his response to them. This is visible in the way he took on preparation making for the Biodynamic Association of New Zealand as a 'job that needs doing'. The responsibility of doing it for others makes him particularly careful regarding the quality of ingredients he chooses (for example, of quartz crystals). His altruistic outlook is also apparent in the way he developed Poraiti farm, took the needs of the Hohepa community into account, and felt satisfaction in seeing the residents' joy on completing a preparation-related task. Chris Hull does not seem to have a personal agenda regarding work with the preparations, and seems to be motivated by his understanding of their spiritual significance.

14. Colin Ross and Wendy Tillman, Seresin Estate, New Zealand

Johanna Schönfelder, Dr Ambra Sedlmayr

Introduction

Seresin Estate is a diverse biodynamic vineyard. It is one of the most prominent places for preparation work in New Zealand, since it actively involves other biodynamic farmers as well as interested individuals from the region in all stages of preparation work. Various events related to the preparations take place throughout the year. All the biodynamic preparations are produced on the farm, and shared with other biodynamic farmers in the region.

Michael Seresin, a famous cinematographer, owns the Estate and employed Colin Ross as his Estate manager from 2006 until mid 2015 when he was replaced by Jared Connoly. Colin converted the farm from organic to biodynamic and took on Wendy Tillman in the role of 'preparations manager'.

The Seresin Estate is situated in Marlborough, the famous wine-growing region in the northernmost part of South Island. The estate of 160 ha (395 ac) comprises three plots of land: the Noa vineyard, with the winery in Wairau valley, and the Raupo Creek and Tatou vineyards.

Though one of the sunniest places in New Zealand, Wairau Valley is classified as being in a cool, temperate region. Average annual rainfall ranges from 965mm (38 in) and there is a strong maritime influence. The annual mean temperature is 12.8°C (55°F). In summer temperatures can reach up to 30°C (86°F). During winter and spring, days tend to be warm (around 20°C, 68°F) and nights are cold. Ground frosts can occur on up to 86 days per year. The plots of land belonging to the Seresin Estate have their own wells or have access to a common well.

Dr Ambra Sedlmayr and Johanna Schönfelder visited the farm on March 23 and 24, 2015 at the start of the annual grape harvest. They visited the Noa vineyard on the first day and carried out an in-depth interview with

Wendy. The second day began with a demonstration of the horse-driven preparation-spraying machine. There was an in-depth interview with Colin during a visit to the Raupo Creek vineyard. Data relating to preparation practice was collected during visits to the preparation store and through the various, shorter interviews that were carried out. No preparation work was done during the visit.

Colin Ross and Wendy Tillman.

Farm portrait

The Seresin Estate is named after its owner, Michael Seresin. Michael Seresin was born in 1942 and grew up in Wellington. He left New Zealand in 1966 and settled in Italy where he started his career as a cinematographer. He became a film director and was known for films like 'Fame', 'Angela's Ashes' and 'Harry Potter'. Colin said that Michael was fascinated by the culture of Italy, where wine seemed to be the ever-present hallmark of 'a civilised life'. Whilst in Italy, Michael also got to know the traditional mixed farming system that was so typical of the 'old world'. Having not come across it before, this way of farming inspired him.

In the early nineties, Michael's interest in wine became a strong passion. As he became aware of the quality wines that were starting to emerge from New Zealand, he decided to return and look for a place to start his own winery. In 1993, he bought 69 ha (170 ac) of land in the Wairau Valley – Noa vineyard. Michael's vision was to have a mixed farm with wine production as a way of life. The trademark of the Seresin Estate is Michael

Seresin's hand print. It symbolises the unique qualities created by the caring, manual work of human beings. Michael's vision is of a farm where people work with rigour, joy, energy, and unrelenting curiosity.

Soon after taking on the farm, Michael changed over to organic farming methods and gradually converted to biodynamics. Although the first biodynamic preparation had been applied in 2002, full Demeter certification was only granted in 2010.

The soils of Noa vineyard are mostly of alluvial origin. The core infrastructure of the operation is located on this plot of land – the winery buildings, wine cellar, machinery, place for making and storing the preparations, offices and a common room where farm staff gather for meals and meetings. The staff come in to work and no one lives on the farm permanently.

View towards the mountains from the Noa vineyard, with a natural wetland and creek.

In 2000 and 2001, two further vineyards were purchased: Raupo Creek and Tatou. Raupo Creek covers 76 ha (188 ac) and is situated in the hills beside Omaka Valley. The soil is rich in clay. Tatou vineyard is a 15 ha (37 ac) property with shingle-based alluvial soils. Noa and Raupo vineyards are not monocultures but diverse farms, with olive groves, extensive vegetable patches, natural wetlands, animal pastures and arable land.

Besides the main products of wine and olive oil, the farm produces a number of other foodstuffs for sharing among the farm staff. There are three dairy cows, seven beef cattle, one hundred and twenty sheep, one hundred chickens, four pigs to dig the ground, two working horses to pull the preparation-spraying machine and two goats.

There are about thirty people working on the Seresin Estate including administrative staff, marketing experts, and people to manage the vineyard and winery – a very diverse team from across the world. The wines produced are exported to Australia, Europe and the United States.

People as part of the farm organism

The Seresin Estate is not only about producing wine in a sustainable, mixed farming system. The role people play is highly valued and they are at the centre of the farm. The Seresin Estate's company philosophy states: 'We see the vineyard as a living organism, a space which our workers, animals, wild birds, flowers, worms and insects can share,' and, 'Farming organically and biodynamically means making a commitment to the land, but that commitment only becomes real by investing in the people and animals that work the vines.' The central role played by human beings in the farm organism and in the production of quality products is evident in the decisions and practices of everyday farm life. There is a deep-seated attitude of generosity on the part of the farm – vegetables, meat and eggs are provided free to all employees. Unpaid workers like WWOOFers (Willing Workers on Organic Farms) are not given tasks that primarily serve the economic interests of the estate, but are instead deliberately directed towards taking care of the vegetable gardens and other work that enhances social life. Colin, as the farm manager, has created space and opportunities for the many people on the farm, enabling them to do the work they really enjoy and are passionate about.

First steps into preparation practice

How Colin came to biodynamic agriculture and the Seresin Estate

Colin was born in 1964, in Malaya (former designation of a federation on the malayan peninsula in Southeast-Asia), where his father worked on a rubber plantation and his mother as a teacher. When he was two years old, the family moved to Australia, and from the age of six, the children were 'invited' as Colin put it, 'to consider wine' as they accompanied their parents to wine-tasting events.

As a young man, Colin was a surfer. His surfing life and travels brought him to Indonesia where his interest was stimulated not only by the waves but also by the Balinese approach to rice farming. He also found, and took up, a form of Indonesian martial art called Silat, which for him 'was a great opportunity to integrate the mind, body, and spirit.' He describes traditional Balinese rice farming (Subak) as a system that finds a deep purpose in all aspects of the farming landscape and has a true reverence for water and life. These impressions helped him decide to work 'with the earth' and become a landscape designer in Australia. He also began to learn about Australian aboriginal culture. When he was thirty he found a job as a wine grower in Australia. He came across biodynamics when watching a slide show of a visit to a large wine company in Burgundy, France. He was impressed by the different look of the biodynamic vineyard. His practical knowledge in viticulture developed as did his career, and for ten years he worked as the estate manager of Brookland Valley Vineyard in Margaret River – a famous Australian wine region. He implemented the estate's conversion to biodynamics.

When he was asked to become the estate manager of the Seresin Estate there were many good reasons for taking up the job. The 'constellation' felt just right to him, he remembers, and he found in himself the rigour, joy, energy and curiosity needed in order to endorse the philosophy and lifestyle proposed for the estate by Michael Seresin. He says of the Seresin Estate: 'The Estate's staff are continually striving to improve all aspects of our produce. We all enjoy the fine Seresin wines and olive oil, plus our own eggs, beef, lamb, apples ... and much more. This has always been part of a dream existence for me.' Colin joined the team at the beginning of 2006.

After he had been managing the Seresin Estate for over a year he found himself having to find some new staff, so he looked for people with a biodynamic mindset and who 'wanted to be more than tractor drivers' – that was when Wendy Tillman was invited to join the team.

How Wendy came to biodynamic agriculture and the Seresin Estate

Wendy was born in 1973, in the United States. At the age of thirty, after some ten years in her career as a project manager for IBM (International Business Machines) in the US, and without any agricultural background, Wendy realised that she wanted to do something that 'makes a difference in the world'. So, in 2005, she took a sabbatical, and while traveling and WWOOFing, came to work on a vineyard in Australia for just one week. Whilst there, she felt a connection with the vine plant on an emotional level, it responded to her and seemed very much alive. She could not really make sense of this feeling, but it was unquestionably real and so she decided to explore it further. She also realised that wine growing was something she could do – 'for there was no reason I could not'. It was during this sabbatical that she came to New Zealand and the Seresin Estate for the first time and 'just fell in love with it'.

When she returned to the US she rejoined IBM on a part-time basis and studied horticulture at a university in her spare time. She then took preparatory courses in order to enter a viticulture programme and finally did a Masters of Science in Viticulture at the University of Adelaide in Australia. Whilst doing her Masters, Wendy attended a workshop on biodynamic viticulture and learned about the basic principles of biodynamics. This workshop was a turning point in Wendy's life. She remembers: 'I loved that workshop. I realised that's what I wanted to do. Growing grapes any other way didn't make sense to me anymore. It was like a switch in my head. From that moment on I knew that I wanted to go into biodynamics.'

The aspect that grabbed her interest was the biodynamic way of looking at everything as being 'alive'. For that was the way she herself looked at the vine: 'I had a real connection with the living vine, the way it grows so quickly and responds to you, that aliveness and the retention of that aliveness. With wine, you're really trying to grow a high-quality product and you want something that is really connected to the earth. Some of the conventional things that are done really just disconnect the plant from the earth.'

Wendy had started reading biodynamic articles and journals and visited biodynamic vineyards in Australia, but still had no direct experience of farming herself. It was more than two years after her visit to Seresin that Colin got in touch and invited her to become the 'preparations manager' at Seresin. Preparations were already being made, but not on a big scale or in a structured way. Colin wanted the biodynamic work to flourish and hired Wendy so that she could coordinate the production and application of the preparations.

Wendy explained that she spends most of her time on the organisational and maintenance tasks connected with the preparations. She said: 'I often work with them from an administrative perspective.' There is a lot of planning needed when working with the preparations on a large scale, as well as actually using them. Her work includes organising the preparation-making days, obtaining the ingredients and organs for making them, checking the preparations in storage (for example, maintaining humidity levels and keeping rats out), keeping records for both Seresin and the Demeter certification (in New Zealand 'preparation biographies' for each preparation have to be produced for inspection by the Demeter organisation), and for planning the large-scale application of the field sprays.

How the work developed

What first motivated Colin to work with the preparations was 'the opportunity to get the farm going without chemicals.' He explained: 'I have had the chance to spray insecticide … to put on masks, gloves, hoods, and suits. I know how it smells, I know the fear in it. I know that it is killing things; I've seen the effect and the result of it… and it doesn't make me feel like a very good human being.' Even though some substances are still used in biodynamic viticulture, like copper and sulphur, Colin is able to use less and less over time as the farm becomes healthier. An ever present motivation in Colin's work is to develop himself in tandem with the farm organism.

Colin first learned about the preparations during a workshop led by Peter Proctor in Margaret River in 2002. Then, in 2007, he attended the biodynamic training course offered by Taruna College in Hawkes Bay. He was by then already working for the Seresin Estate. He was deeply impressed by the explanations given by Peter Proctor during the workshops about the 'information streaming in from the cosmos' and the way he could speak about it in such simple and common language. That something so spectacular to Colin was a matter of daily reality for Peter Proctor, touched him deeply.

For Colin, it was not reading about them but actually working with the preparations that sensitised him to their special qualities. 'There is no substitute for doing it, for seeing it,' he said. 'It's a wonderful, amazing, powerful feeling when you're all part of making, applying or lifting the horn manure. There is a community, it feels very true, real and connected.' Looking back over the first three to four years of dealing with the preparations, Colin remembers how he struggled to get the work done. There were always challenges to getting everything organised and operating well on a technical level. But 'I think the more we do it, the easier it becomes. It becomes second nature and part of the language, it's not something you have to rediscover. So yeah, it's getting easier and easier.'

When Wendy started her job at the Seresin Estate she did not have any formal training in making the preparations. She had only read Peter Proctor's book *Grasp the Nettle*, which to her is 'the New Zealand preparation-making guide and a good book for getting started. I haven't actually seen any other book in English,' she said, 'that really gives you the basics of how you make them and how you store them.' But even though she sees Peter Proctor as 'the greatest authority on preparation making in

New Zealand,' Wendy says: 'I don't do everything that he says either. I listen to it and take on board what works for me in my situation.'

So from the beginning that book was her guide and Colin was her teacher. 'Most of what I learned, I learned from him,' said Wendy. Her work with the preparations has gradually evolved during the course of her seven years working for the Seresin Estate. This development has been more about 'fine tuning' and she experiences it as a gradual process of making 'little bits of improvement to your understanding.'

Establishing a cow herd to provide manure

The first preparation Wendy ever made herself was the horn manure preparation (500). Her first big project as 'preparations manager' was to find a suitable source of cow manure for making the horn manure preparation. Not content with the manure of their own beef cows and the quality of manure from neighbouring farms, she soon came to the conclusion that, as a biodynamic farm, the Seresin Estate needed its own dairy cows. She had never handled a cow before, but that did not stop her from buying a dairy cow with its calf and creating what was needed to look after them. Wendy designed a dairy where the cow could be milked and the milk stored. Considering herself a 'naive beginner', she just followed her own sense of what would suit the needs of the cow and the place. Finally, in 2014, the Seresin Estate was able to use the manure of a lactating cow that had been born on the property for producing horn manure preparation.

Changes in response to practical challenges

Alongside Peter Proctor's book and Colin's teaching, Wendy was able to advance her skills and understanding of the preparations by attending biodynamic and preparation-making workshops, and by reflecting on the questions asked of her by visitors to the Seresin Estate. Some practices were modified in response to mishaps, such as the complete loss one year of the buried chamomile. Wendy had previously buried the stuffed intestines in a single clay pot with a lid, but when she unearthed the pot, no sign of a preparation could be found. It was 'as if it (the chamomile) had not existed.' She now divides the stuffed intestines between two separate pots for burying, puts a plug in the bottom of the pots and seals the rim of their lids with clay from the vineyard. She feels that the preparation is now more secure. Wendy has found that the preparations are prone to disappear in the local soils and so prefers to place them in the soil in clay pots.

Wendy allows the filled chamomile sausages and the dandelion mesentery packages to dry a little before burying them. Someone had recommended this practice in order to 'give them (the assembled organs and plant materials) a little bit of time to become their own thing, and to cure a bit.' This feels right to Wendy and she has the impression that it allows the preparation to 'get used to itself before being put into the ground.' There is also another more pragmatic reason for this procedure – a lot of preparations have to be made on such a large estate and it is easier to manage the work if it is spread out over two days: one day for assembling the preparations, and another day, later in the week, for burying them.

Changes guided by observation and intuition

When it comes to making changes in her work with the preparations, Wendy evaluates the questions raised and the advice given to her by drawing on her own observations and feelings. 'One of the things the practice of biodynamics has taught me,' explained Wendy, 'is the need to observe – using the power of observation to discern what needs to be done and then following that intuition. I think that I do listen to what people say about the preparations, the different advice, and different ideas. But in terms of 'Is that the right thing for me to do for this place?' it (making decisions) is definitely more intuitive and based on my own experiences, observations, and feelings.' Wendy gave a specific example of the way her work with the preparations has been adapted. She recounted how, based on her intuition, she introduced changes to the way CPP is made: 'When we made our Cow Pat Pit preparation, we used to take a portion of mature Cow Pat Pit, put the preparations into that, and then mix it into the new Cow Pat Pit. Someone once asked me: 'Do you think that's the right thing to do?'. I had never thought about it, it just seemed the right thing to do, almost like inoculating the new material with a bit of the old. It seemed to make sense from a microbiological perspective but perhaps not from a living perspective. The more I thought about it, the more I felt that the Cow Pat Pit was its own living entity and that I should not force a different entity upon it.' Wendy, therefore, stopped inoculating the mixture with mature CPP and instead started applying the preparations directly.

Wendy says that she is still learning 'to ensure that my time with the preparations is both focused and dedicated'. She explains that 'Because I do so many different things in my job (and I'm always doing lots of things), it sometimes happens that when I work with the preparations I have other things on my mind, and because I have so much else to do, I tend to rush things. It feels different and much better if I can allocate myself a good stretch of time and create the space to really focus on the preparations, to let go of all the other things in my head. Working with the preparations does feel very different from driving a tractor or washing dishes or whatever … They feel alive, they command and deserve our full respect, focus, and dedication.'

Colin's and Wendy's understanding of the preparations

Colin's understanding of the power of intentions

Having long been a surfer and martial arts practitioner, Colin knows how powerful intentions can be. This experience is very important to him. 'If you're going to paddle and catch a wave, you've gotta think really clearly that you're gonna have to go faster than gravity and become sort of at one with the pace of moving water. And if you're hesitating and pull back, it doesn't work. You actually have to throw yourself into it. That makes sense. So your intention has to be very positive about what you're gonna do.' He added: 'Martial arts is the same, you have to train in order to bring your intention into action. So for me that's my interpretation of "dynamic".' Transferring this understanding to biodynamics, he says: 'In practical terms it means how you set about stirring. You should have the intention that you're doing it with rigour, doing it with joy, doing it with energy and, yes, even doing it with unrelenting curiosity too. It means you're involved, you're observing it, you've got some dynamic, you're moving.'

Colin believes that not only humans, but nature too can bring intention into manifestation. He refers to Steiner: 'I love this thought that "matter is never without spirit and spirit is never without matter." You can take that into everything, it's so simple. We spend a lot of time talking about intention: the ability to bring intention and have it manifested in matter. I imagine, I think, I create, I do.' In Colin's eyes the beings of nature also bring intentions to manifestation. What nature has once produced, can be produced more easily a second time: the beings of nature have the capacity to learn and to remember. His understanding is that nature can reproduce and recall everything that it has once learned and experienced. From this point of view, he endorses the belief held by Australian aborigines that the land itself has memory. Colin is convinced 'that there is intelligence not only in the vine, but in the soil too and I'm sure it's in the plants.'

Colin's understanding and work with the intelligence and the memory of the land

For Colin, nature's intelligence is expressed in the way it is so perfectly organised. Colin experiences the Earth as a living organism and nature as a place of 'perfect order'. He recounted that: 'I've spent a lot of time in old growth forests … there is a perfect harmony. If you go swimming on a coral reef no single piece is out of place, it's all perfect. So behind what we see there is an order in nature.' This capacity of nature for self-organisation can be built on when it comes to farming. To manage the farm as an organism gradually enhances its ability to organise itself and attain an ever greater balance. In Colin's experience, 'the further we proceed down the biodynamic route the more we find a different balance in the landscape as well.'

When observing the estate from a helicopter he could see shafts of light shining across the plots of land belonging to the Seresin Estate. For Colin, this was evidence of the farm being as an organism and suggests, 'that it could be an observable phenomenon.' He recounted: 'We use helicopters to fight frost sometimes. And once we had a group of sommeliers … and we said: "Why don't we take these guys, fly over the vineyard and show them where we are." So we flew around and when we came closer we could see what seemed like a haze over the land – like a shimmering light. And you could see from afar three distinct shafts of light, and as you flew up you could see it. It wasn't just me who saw it but three others saw it independently. It was not just a bright light, it was a lifting, rising light. We didn't talk in the helicopter. But when we got out, everyone went: "Did you see that? Your property stood out kilometres away, and not just the greenness of the grass. There was this light."'

Even though the Seresin Estate is split up over three separate plots of land, Colin experiences a single identity and also tries to maintain the individuality of each plot. This means in practical terms that the woodchips or grape pressings used for compost are used on the land of their origin to enhance individuality. The wine from the different plots is also kept separate, and Colin sees this as a 'message in a bottle'. He explains: 'We can feel the differences in the land. It's all developing in a common direction and so is still probably the same message. But each plot of land has a distinct quality and you would never ever want to lose that.'

The intelligence of the farmer complements the intelligence of the land.

The farmer guides the processes that enable a balanced landscape to develop. Colin manages the farm in order to work with and enhance the intelligence of the land. This is evident, for example, in his approach to pruning. As he explains: 'The shape we prune the vine to is more like a pear tree than a vine (bilateral cordon). Our understanding is that old wood is full of carbohydrates, but I also believe that when we prune in this way … there is naturally more mass available to accumulate that memory, that intelligence.' He explains that in addition: 'The yeasts that do the fermentation, we find that they grow as much on the bark as on the fruit. So having more bark available means more habitats, more homes, for all these micro-organisms.'

The wine provides Colin with feedback on the management of the whole farm. When he looks at the conventional vineyards in the neighbourhood, he sees that they are indeed ordered, quiet and clean, but they are also places with very little diversity, with only two species growing: vines and grass. He expects wines produced in this way to be 'very fruity but not of interest,' for 'the vine behind it is very mechanical.' He explains that conventional wine growing separates the vine from the landscape and therefore disconnects the plant from 'the intelligence of the land'. He explains that were he to make wine from conventionally grown grapes, using only the yeasts on the grapes, it 'wouldn't make itself into wine' because there aren't enough natural yeasts living on the conventionally grown vine. In order to enhance the land's 'memory', Colin tries to accumulate and return the memories and experiences to it. One way of doing this is to spray the lees (dead yeast deposit from wine production) on the vines so that they learn about the making of wine too. He explains: 'We spray this back on to the trunks to really close the cycle. We're trying to put the memory back.'

If the Earth is seen as a dead, inanimate object, it will be farmed accordingly, asserts Colin. But if the Earth is conceived as a living being, then one needs to ask oneself: 'Is this action going to enhance or diminish life?' Colin says that bringing more life to the Earth is the 'first intention of his actions'. This also means for Colin that 'I'm not going to try to control powdery mildew by applying 501. I'm not going to try and control the grass grub beetle with 500. To do so means your mind is already thinking of the "dis-ease". What then is the state of "ease"? When everything is full of life. When do you feel best? When you feel lively. So the question is: What is your intention when you work with the preparations? It's about bringing more life to the land.' Thus a fundamental principle and motto on the Seresin Estate is to create 'farm ease and not fight dis-ease'.

Wendy's approach to understanding and working with the preparations

Although Wendy has a scientific background, she feels no personal need for a scientific proof concerning the working and effectiveness of the preparations. In her eyes natural scientific explanations of the preparations miss the point of what biodynamics is about. She explains how: 'You sometimes meet people who really want to challenge you about what scientific evidence there is for the preparations. I try to learn enough about the preparations in order to be able to make basic sense of them in terms of their connection to calcium and silica, clay and such like. But I don't need it proven from a scientific perspective, because I see the evidence. And I almost feel that people who need it scientifically proven to them are missing the point. They're missing the point of biodynamics. There is a bit of trust involved in all of it.'

Wendy trusts and follows her intuition with regard to the preparations. The positive results are evidence enough for her. She explains: 'I have let go of trying to understand them and just trust that they work and that they work together. The biggest thing for me is seeing the results and seeing that it actually does work: seeing the changes on the land and the amazing food and the amazing wine that we can produce, and seeing the difference between our land and our practices and those of the people around us.'

Wendy described how she always tries to listen to her intuition and that it is intuition that leads her through life. The use of the preparations feels intuitively right to her and it is 'one of those things that I just have to listen to, that I feel works and is important for the system and the health of the land.' With regard to the practical aspects of her work she is guided by the 'logical' side – in her role as coordinator she has to organise all the practical aspects of working with the preparations.

Wendy is happy that her work creates a place and products that people can enjoy. Wendy thinks that not only the staff, but also people from around the world who drink Seresin wine, can benefit from the care given to the land and from the enhanced life forces contained in the products, due to the application of the preparations. At the same time she feels that her work fulfils her own needs: she loves the combination of structuring practical work using her logical and organisational skills, while the work itself – like picking flowers – nurtures the soul. Working with the preparations always puts Wendy 'in a good mood'. She says: 'I'm not quite sure why I enjoy doing it so much or why it comes so naturally. It's one of these things. When I went to my first biodynamic workshop, it made such sense to me, it completely changed my life and the path I was on.'

> ### *An impression of the interview situation*
> Johanna Schönfelder
>
> Since in this interview Dr Sedlmayr was the leading interviewer, I was able to not only listen to the dialogue and keep track of our questions, but also observe Wendy's mood and follow how the interview flowed. I enjoyed how the intuitiveness of Wendy's words led directly to a coherent feeling. The answer to one question led to the next question. The immediacy of Wendy's words combined with her alert and open-minded nature made it easy for me to not only hear about her approach to the preparations – but directly to 'plunge' into it. I found it very refreshing to hear someone originally coming from a scientific background talking in such a straightforward way about classically 'unscientific' topics. Wendy's passion, positivity and mercurial mind really inspired me.

Making sense of the preparations

For Colin, wine has the ability to express everything that affects the plant's growth – be it the characteristics of the growing season, human interventions and especially the soil. He has a detailed understanding of how the quality of the soil is expressed in the final product. Within this 'the preparations help to ground the plant, connect it to the soil.'

Making sense of horn manure

Making sense of horn manure is for Colin an easy one for when he unearths the horns he finds them covered in living substances. He sees this as a picture of what farmers can do with the horn manure preparation – activate the living part of the soil. He described how: 'You put the manure inside, and it comes out looking totally different. It does not require a great leap of faith to perceive that. We all know, we understand how earth smells, how soil smells. We understand what humus is, we understand that the only thing we can really do as farmers is activate that living portion of the soil. You can easily make a connection between what you see and feel coming out of the horn and what you can feel and see in the soil. So it makes a lot of sense to me.' Spraying the 500 to Colin feels like 'you've got a bucket full of living earth.' Using horn manure preparation makes him feel 'cooler' like 'it's sort of going into the night.'

Making sense of horn silica

Having a concept of horn silica is far less tangible for Colin. Spraying the 501 causes technical problems because tiny pieces of wood scraped off the barrel while stirring can block the mechanical sprayer. 'I'm not attached to the 501,' said Colin. He experiences less of a biological process with this preparation and this makes it harder for him to get a feeling for it. Making horn silica is a noisy and dusty process and even the stirring feels 'cold' to him for it is mostly done with machines. But he can connect to the preparation when it is being sprayed out. Colin describes how the horn silica preparation causes a bright and upright feeling in him, and makes him 'more aware of the change of the day'. He also observes how the preparation dries and hardens his skin when it comes in contact with it. It feels to him like 'life spraying', a really 'powerful thing'.

Nonetheless, Colin feels that the 501 is the preparation of our time and says: 'I think a lot of people see it in terms of enhancing sunlight capacity, but I like the idea of it being like little memory chips that are sprayed out. The physical effect of silica is to harden the cells and increase sugar levels, but it also feels as though it is connecting everything together to a greater extent than we think. We had a discussion here the other day. There was once the age of wood and we had the wood age people, then Stone Age people, then Bronze Age and Iron Age and so what do we have today? One idea was that we are in the silica age. So I think that horn silica is also a preparation belonging to our times.'

Making sense of how 500 and 501 work together

Colin is convinced that the two field spray preparations work strongly together. He experiences them as the link connecting soil and cosmos. He described the two polarities: 'We know that the plant lives in the soil, it has roots, we know it grows in the air. You know about the yin-yang symbol? In Tai Chi you're grounded on the earth and opened up to the cosmos as well.' As Colin sees it, the line dividing heaven and earth is undefined and variable. A new element combining the two polarities can enter and dissolve the boundary. Colin describes the space where heaven and earth come together in the human being as follows: 'Don't think, just do it. I think that in training your awareness you learn how to let go of your mind. Where does an answer and all that stuff come from? How to stop thinking and start trusting?' Referring back to the question he was asked he said: 'You're asking how do we know if these two forces work? You're dealing with matter and you're dealing with space.' He seems to be referring to horn manure as a representative of matter and horn silica as a representative of space and that the two interact naturally with one another to allow new life, or a new ordering of life, to emerge.

The effects of the biodynamic preparations

General effects

Both Colin and Wendy have observed changes on the Seresin Estate that they attribute to biodynamic practices. Colin stresses that it is difficult to single out the effects that have been brought about by the biodynamic preparations themselves, and what effects are due to their steadily improving growing experience, other biodynamic practices or even environmental conditions (including beneficial cosmic constellations).

Both Colin and Wendy have seen steady improvements at the Seresin Estate in terms of the quality of the top soil and improved vine health. Colin has also observed that the undergrowth has become less grassy and less invasive, and is now composed of a diversity of herbs. Wendy's observations focus on health, balance and quality aspects. She feels that biodiversity has increased and that both plants and animals have become more healthy, balanced and vibrant. This expresses itself for her in the quality of the vegetables and wines produced.

Colin also considers that the preparations have an effect on the people working with them, since, 'I don't think you can apply the preparations without growing more curious about the soil's mineral content. You can't apply the preparations without becoming more curious about the biology of your soil. So the two – horn manure and horn silica preparation – are not just working on the land but on you as well.'

Effects attributed to the field spray preparations

Even though Colin is cautious about establishing direct cause-and-effect relationships in connection with the preparations, he explained his understanding of the way the horn manure and horn silica preparations work. Colin believes that horn manure helps to balance a plant's physical and etheric body. Plants that are less stressed are less prone to being attacked by pests. Colin has observed this, for example, in the reduction of problems caused by chewing insects such as grass grubs. From his point of view, horn manure preparation improves balance in the whole landscape.

Colin perceives a relationship between the application of horn silica preparation and the ripening process of the grapes. He said: 'There is definitely a physiological thing going on. We've seen how the physiology of the plant has changed so that we're getting ripeness at lower sugar levels.' Colin believes that the horn silica preparation plays a role in reducing acidity levels and ripening the grapes. He explained his complex picture in the following way: 'What is ripeness? When the acid is in balance with the sugar, it's ripe. If we can get the acid to go away… Everybody knows where sugar comes from, but where does acid go? To my understanding, the more I look at it, the acid is being fed to the bacteria in the soil. So a more active biology in the soil, maybe that's helping, the vine is excreting as well.' The idea that plants feed the soil life underlies several management decisions. Colin said: 'We shouldn't be harvesting too early because we have to feed the rest of the environment. We know that the plants have a relationship with the biology in the soil. What do they eat? We try not to prune until after the winter solstice because until then the vine is still excreting. The sun has to get into the earth somehow. So it's part of that journey.' In summary, the main effects resulting from this complex process involving horn silica preparation are that the brix level, the measure of the amount of dissolved solution in a liquid, is raised, grapes ferment more consistently (less erratic fermentation), and they seem more resilient against fungal attack even after heavy rains.

Effects attributed to the compost preparations

Colin considers the compost preparations not as separate elements, but as functioning collectively as one organ. His experience of applying the compost preparations is: 'Of giving a sense of shape to a more conscious living being, the compost becomes a living entity.' Having put the preparations into the heap it 'feels as though you've completed the process and breathed life into something.' To his understanding, the task of the compost preparations is to foster life by helping to bring earth and sky together. He described it like this: 'Earth and the sky talk to each other – through life. We would generally just use the Cow Pat Pit (we've had the preparations in it) and see how the life force of everything is increased.'

Quality aspects

Wendy is the person at Seresin who generally checks and reflects on the quality of the preparations that have been produced. Her experience with the preparations has led her to develop different quality criteria for each preparation based on her sensory perception. She described how: 'I look at slightly different things, depending on which preparation it is and how they were prepared. I will look differently at the nettle than I would at something that's been in an animal sheath.' Taking the nettle as a practical example she describes: 'In terms of quality what I like to see at the point of harvest is a moist but not wet preparation, with something of an earthy or plant odour, but no anaerobic smell to it. When it comes out I like to see it without a whole lot of the texture of the leaf left, but rather like a black, sandy substance ... it shouldn't cohere, it shouldn't stick together.'

It is important for both Wendy and Colin that the buried preparations do not become anaerobic, since in their opinion this would negatively affect the quality of the final product.

The social setting

Preparation work as a social occasion

Work with the preparations on the Seresin Estate is consciously made into a social extent. Wendy explains that making and applying the preparations is 'an occasion for coming together as a community and catching up with people from the region.' The idea of stirring the field spray preparations as a collective effort, for example, is based on the view that good social interaction and joy make a positive contribution to the preparation and ultimately to the wine. Wendy said: 'When we work with the vines, the land and with everything that we do, we're trying to give our best in terms of our energy and care. We're doing things thoughtfully and calmly. When we're stirring the preparation, we want to give the preparation all of our good intentions so that all of our care and concern can be transferred to the plant.' Wendy describes this collective stirring and spraying as being 'almost like a celebration ... There's lots of joking and talking, people walking back and forward between barrels, and stirring. And after we've finished spraying, we all come back together and say, "Thank you," and congratulate ourselves for a job well done.' At the same time there is a clear purpose, a 'definitive intention' in doing this work together: 'Colin always says something before we start, to talk about why we're here and what we're trying to do.'

All the vineyard staff, people from the winery and office as well as biodynamic farmers from the region, are invited to help with this practical work with the preparations. It is particularly when making the compost preparations in the autumn that ten to twenty biodynamic farmers from the region come to Seresin, using this occasion for a visit. Whereas some biodynamic farmers in the area make their own CPP and horn manure preparation, it is rare for them to produce the compost preparations on their own farm.

The dates of making and applying preparations are publicised and people who are interested can come along and join in. The tasks shared include making and applying horn manure preparation, making compost preparations and inoculating the compost heaps. For example, the horn manure preparation is sprayed by some thirty to fifty people using brushes or branches. For Wendy, so many participants require that someone coordinates the work. She said: 'A little bit of randomness is good, but you need a structure to go with it, otherwise you just have to do it all again the next day.' Since the horn silica preparation is sprayed in the early morning using a fine mist over larger areas, it is done with tractors.

With regard to connecting biodynamic famers, Wendy explained that the preparations help contacts to be maintained throughout the year: 'I generally keep the preparations here. So when someone needs them, they come here and say: "We're making a Cow Pat Pit, can I come and get some preparations?" This encourages ongoing communication throughout the year.'

Colin says that he doesn't feel isolated with his biodynamic thoughts and practices. There is a national Biodynamic Association conference once a year and also meetings of an organic wine growers group that includes biodynamic farmers. Colin reads magazines from Australia and New Zealand to get a broader look on biodynamic topics and gets feedback through the annual Demeter inspection. The group of organic wine growers feels like a collective – The organic wine industry in New Zealand is small, and each one of the Demeter producers knows each other.

The quality of the wine is the best argument against critics

Asked about sceptical people and the 'exposed' position of a biodynamic farm in a neighbourhood of mostly conventionally run farms, Wendy explains: 'I'm sure there are critical people, but we don't tend to have them come along too much. I think most of the people that come do so because they are interested and intrigued and want to learn more, and not because they're trying to talk themselves in or out of biodynamics. I think Seresin has been doing organics and biodynamics in this region longer than anybody else and we're pretty big. I think in the beginning people had a different perception, people in the broader community here. I think they thought a bit more like – we're a bunch of hippies, they have weeds everywhere because they don't mow everything. But I think now that you can really see the results in the wine and how amazing the fruit is, I think that sort of thing has made people think: "Maybe what they're doing makes sense." At least they can't argue with you, because the wine is fantastic. If it works for us, it might not work for them, but it still works for us.'

Preparation practice

Apart from the eight classical preparations, Cow Pat Pit, horsetail tea and compost teas are also made on the Seresin Estate. Most of the required ingredients come directly from the farm. While oak bark comes from a another farmer, dandelion and yarrow can be picked all around the fields of the Seresin Estate. Chamomile, nettle and valerian are cultivated and picked in a small garden.

The intestine and mesentery needed for making the preparations are taken from beef cattle, which are shot to provide meat for the staff, using the so called 'home kill system' that allows an authorised person to kill animals on the property. Since the skulls of the cows get broken when shot, the skulls for making oak bark preparation are taken from sheep.

Field spray preparations
Obtaining and handling horns
At Seresin, the number of horns needed is much more than the farm can produce. Fortunately, they have been able to establish a connection with a regional home-kill butcher who gathers horns for Seresin. The horns obtained from him are mostly non-organic and come from steers as well as cows. It is very difficult to get the 'ideal' horns from biodynamic cows. The horns that are brought by the butcher are put into a worm-farm to remove the core. The horns can be used for more than five years.

Horn manure preparation (500)
The little milking cow herd at the Seresin Estate consists of three animals. It takes about a week to collect enough manure from them to fill the 800 horns needed for the whole estate and wider community. During the period when the manure is being collected, the grazing area of a lactating cow is restricted and supplemented by hay to ensure the dung is firm enough. The horns are filled in mid-March before the grape harvest and ideally during a descending moon period and in an earth sign (Virgo, Capricorn or Taurus). They are buried in a circle with the openings towards the soil so they do not fill with water. Before the whole batch is taken out, sometime after the spring equinox on September 22, Wendy digs out some horns to check if the horn manure

preparation is ready. Based on these samples a decision is made as to whether the horns should be unearthed or left in the ground a little longer.

Horn silica preparation (501)

The quartz crystals used for producing horn silica preparation are bought on the internet from an established crystal dealer. They are sourced in South America. This is because local crystals are considered to be 'too cloudy'. A steel mortar has been specially built to smash the stones. The smashed crystals are sieved and then rubbed to powder between glass panes. Seresin staff and members of the community are invited to help with the grinding work. The powder is mixed with water to create a paste and then spooned into the horns. The paste settles down and the horns are buried with their tips pointing up towards the cosmos. The horns are filled in early September and they are unearthed in autumn (March, April) during an ascending moon period in a fire sign (Aries, Sagittarius, Leo). The preparation is stored in glass jars on a window ledge in the vineyard office.

The horns are buried in a circle.

Stirring and applying the spray preparations

At the Seresin Estate the field spray preparations are either stirred by hand, for smaller amounts, or with a mechanical stirring machine and a flow form for bigger amounts. The water comes from a well and is heated up using a boiler or gas burner. Spraying is either done by hand, using the specially constructed horse-sprayer, or using tractors.

The horse-drawn spraying machine is only used to spray the 501, and the horses cover only a small portion of the total area.

The horn manure preparation is mostly stirred by hand in old wine barrels using a stick similar to a hard broom hanging down from a beam with shorter sticks attached to the end. Depending on how much is needed, some of the horn manure preparation is also stirred using a flow form or a machine. About 50 g (1¾ oz) of horn manure, stirred in 32 l (8 ga) of water is needed for one hectare (2½ ac). It is sprayed once in spring when the vine buds burst. About 100 g (3½ oz) of CPP is added during the last 20 mins of stirring.

Hand stirring is usually done with three people standing beside each barrel. Each one stirs for about 5 mins and then the next person takes over until an hour of stirring is completed. The idea behind this is to share the work out and make sure there is not too much work for one person, but 'also to help keep the dynamic going, the flow going, the conversation going, making it a bit more social. So that we're not standing doing nothing, we're all participating in it and taking turns.'

Since 3,000 l (780 ga) are needed for a single application, the bulk of horn silica preparation is stirred using either the flow form or stirring machine. 2.5 g (½ tsp) are used in 32 l (8 ga) of water per hectare (2½ ac). In order to focus more human care and attention on the silica preparation, a group of people always stir one barrel by hand and add it to the horn silica that has been stirred by the machine or flow form. Wendy explains: 'We stand next to the machine and do one barrel by hand. We all stand around and take turns stirring and talking in the morning. And then we take a bucketful out of the barrel and put some of the hand-stirred preparation into each of the tractor sprayers.' The idea behind this is 'that it's not just some mindless activity of pumping it from one machine to another machine, but that we've actually stood there, talked about it while we're stirring, have had that time together and are then passing all that energy and intention into the product.'

Spraying is ideally carried out 48 hours before full moon in order to counter the moist lunar influences that can lead to the development of mildew. The preparation is first applied to the vines when the grapes have reached pea-size and on further occasions up until harvest, depending on seasonal conditions.

The experience of stirring and spraying

Wendy experiences stirring the 500 and the 501 quite differently. For her, there is a sense of 'aliveness' with the horn manure preparation, something that is also reflected in the active social involvement of people while stirring. The horn silica preparation by contrast, is of a mineral nature and according to Wendy 'seems much colder and, I guess, more inert. Not quite alive. It has energy in it, but a different type of energy than the 500. It's more of a quiet activity. I think this is partly because it's so early in the morning. Everybody has this calm stillness of the morning, it's not big, jovial laughing, sharing stories over the preparations. It's much quieter.'

Since Wendy also stirs by herself for her home garden, she can compare the experiences of shared communal stirring with doing it alone. She describes the experience of stirring by herself: 'I find the time goes by really fast. I guess I feel drawn into the centre of the vortex, I find it very meditative.' Her experience of stirring on the farm is always coloured by her role as the coordinator of the work and as she explained: 'I have to work out all of the logistics, so in my mind I'm also thinking: "Did I put that there, do I have enough brushes, where does that thing go?" So my mind is less free, because I do the coordination. At home, my mind can be more free. When we do it as a group, I find it less meditative.' When stirring with other people, Wendy sees her role as bringing the intentions of the various people together – quite opposite to the experience of stirring by herself: 'It's not about me, it's about everything around me and bringing all of our intentions together. I find it more of a social reaching out to people as opposed to the stirring drawing me into the preparation.'

The horse-drawn spraying machine.

The horse-drawn spraying machine at work.

Mechanical or manual spraying

For Wendy, there is a big difference between spraying by hand or using spraying equipment attached to the tractor. Using the machine is 'not the absolute perfect way to do it', but since it is not possible to spray everything out by hand it is 'the next best way'. Wendy feels she has to compensate for using the tractor and so, while spraying with it, she tries 'to think about what I'm doing, why I'm doing it and keep that in mind while I'm doing it. Hopefully that thought and that energy will undo some of the effects of doing it with a tractor. My ideas and my intention are there even though I have a piece of machinery between me and the preparations.' She explains that: 'Spraying by hand feels very freeing and very like a loving activity. You really feel, when you're flicking the 500, that you're spraying little drops of joy everywhere and you feel really connected to your environment. Flying back in your face as the wind picks it up or as you walk on the grass you see it landing at your feet. I think that you're so much more connected to the environment and it's a much more joyful experience than doing it with a tractor. Because there are also a lot of other people around you, there is again lots of laughing, joking, talking – that social aspect that you're imbuing the vines or the land with, that happiness. And it's probably a bit calmer, a calmer feeling. You're walking at a pace that's comfortable to you, whereas driving a tractor you go quite quickly and you're being driven by a machine.'

Compost preparations

Wendy tries to produce 300–500 sets of compost preparations each year. One set contains 1 g (¼ tsp) of each preparation. It is not all used on the farm; some is given away to people for their own compost or to make Cow Pat Pit preparation.

During autumn – March and April – is when most of the compost preparations are made and placed in the earth, and they are unearthed after the spring equinox in September and October. The horn silica preparation (501) follows the reverse rhythm, and the nettle preparation (504) is buried and lifted in March, staying in the ground for one whole year.

To protect the preparations from rats, being washed away or 'mysteriously disappearing', none of them is buried directly in the soil. Instead they are all placed in a clay pipe or pot for protection. This practice also simplifies things when it comes to lifting them out of the ground. This is not done as a group event but is carried out by a small team lead by Wendy, who monitors their quality on the spot.

Yarrow preparation (502)

Yarrow grows wild in New Zealand and can be picked all around the Seresin Estate. It is only the individual flower heads and not the whole corymb which is used. The flower heads are dried in an airing cupboard. The stag's bladder comes from a deer farm and is stored dry until used. The stag's bladder is stuffed in springtime with the yarrow flower heads that have been moistened with yarrow tea. It is then left to hang in a tree over the summer and in the autumn it is put into a clay pot filled with compost and buried.

Chamomile preparation (503)

The intestines used are either obtained fresh or frozen. Before filling them, they are cut into lengths that can fit into the clay pots used to protect them during their time in the soil. The dried chamomile is moistened with some fresh chamomile-flower tea prior to filling the intestines – a process facilitated by the use of funnels and sticks. The preparation is then hung up and allowed to 'cure' for two to three days to enable the organ and plant material to become one entity before being given over to the soil.

Nettle preparation (504)

Urtica dioica does not grow wild in New Zealand, but it has been grown from seed and a large patch of nettle is now readily available. The leaves are collected and dried in December and January and then pressed into a clay pipe, which is then sealed with stones and clay. The nettle preparation is made and buried in March and dug out one year later.

Oak bark preparation (505)

Sheep skulls are used. The brains are flushed out with water, but the flesh is left on the heads. The oak bark is rasped off a tree, moistened with oak-bark tea and then put into the sheep skulls using a funnel. The hole is plugged with willow. The skulls are then placed in a barrel of water containing decaying vegetation.

Dandelion preparation (506)

Sections of greater omentum taken from one of the Seresin's cows is either used fresh or frozen, in which case it is defrosted and put out in the sun to soften. The dried flowers are moistened with some fresh dandelion-flower tea. They are wrapped into packages and tied together with flax string.

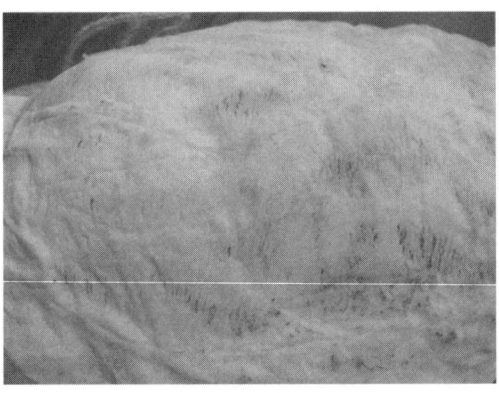

A cow's greater omentum is used at the Seresin Estate for producing the 506.

Valerian preparation (507)

The florets are picked fresh, preferably on a Flower day. They are crushed and filled into bottles the same day and covered with water – 20% fresh, crushed flowers to 80% water. The mixture is left to stand on a sunny window sill for one week and is then filtered.

Applying the compost preparations

Between 300 and 500 tons of compost is produced on the Seresin Estate each year. Winery waste – mostly grape pulp and lees – is the main component. It is complemented with willow wood chips (a source of renewable carbon), hay collected from the headlands, old mulching material, woody material from native tree plantings, fresh-cut creek trimmings, basalt, rock phosphate and cow dung. As soon as the heap is completed it is inoculated with the compost preparations. Wendy uses one set per 5 m^3 (176 ft^3) of compost. The compost preparations are added by making small balls of CPP, inserting one preparation into each ball, making a hole with a pole and placing the ball deep inside the heap. The valerian preparation (507) is mixed with water (10 ml, ⅓ fl oz, of a 5% solution in 13.5 l, 3½ ga, of water per 5 m^2, 54 ft^2), stirred for 10 mins and then put into holes on the top of the pile as well as sprinkled over the whole pile. Then the holes are covered.

The compost is applied to the land in large amounts using a machine that mixes it with water and sprays it out. Some compost is also sprayed by hand.

Burying and storing practice

Burying practice

The filled sheaths are placed inside clay pots surrounded with compost, and the nettle preparation is buried directly in a clay pipe. The pits they are buried in contain a compost-soil mix. The pots are covered with a lid and sealed with wet clay from the farm. The clay pipes are closed with stones and also sealed with clay. They are buried at a central location within the estate. Wendy explains that she surrounds the preparations with compost in order 'to make a bed for the preparation, otherwise they are completely disconnected from the soil. If there is good quality soil around, I would consider putting this in the pots instead of, or in addition to, the compost.'

Storing the preparations

The preparations are stored within specially designed clay pots – Seresin's hand print trademark is pressed into the lid of the pot. The clay pots are stored safely in old, peat-filled wine barrels. Wendy tries to keep the peat in the barrel moist, and because the clay pot exchanges moisture with the peat the preparations are able to maintain a constant moisture level. This means that Wendy rarely has to add water directly to the preparation. She explained that working with moist peat is the usual practice in New Zealand and is recommended by Peter Proctor. Since peat is hydrophobic, it takes some time before it is fully soaked with water. It took several years for Wendy to achieve the right level of moisture in the peat. Wendy likes the preparation to be slightly moist in the store. She said: 'When I touch the preparation, I want it to feel alive.'

The storage barrels stand on a concrete base in the winery, underground. 'To put them into a living environment would be the ideal,' says Wendy, but when they tried to store the preparations underground outside, rats found a way of getting in. Flooding can also be a problem in the region and could lead to the complete loss of the preparations if they are stored outside. So they decided to store them 'underground but not outside'. Each preparation inside the barrel is marked with a tag showing the name of the preparation and the date when it was lifted.

Preparation storage: specially made clay pots containing the preparations are kept in old wine barrels filled with peat in the wine cellar.

Derived preparations and other applications
Cow Pat Pit preparation (CPP)
At Seresin Estate 700–1,000 kg (¾ – 1 ton) of CPP are produced each year. This has been developed on the advice of Peter Proctor, who was doing consultancy work for the Seresin Estate from 2004 to 2005. His suggestion was to make and use CPP in order to maintain the vitality of the land. Following his instructions, 21 brick pits have been constructed in the soil with a simple roof over them to provide shade. The bricks keep rats away,

invite worms in and help retain moisture. The individual pits are covered with wooden lids.

CPP is made all year round. To produce it, cow dung is collected every second day. The manure is mixed with shovels in a trailer for about one hour. 200 g (7 oz) of ground egg shells (using the egg shells from the farm's own chickens) and 300 g (10½ oz) of basalt dust are added per pit while mixing. Mixing aerates the manure. When mixing is completed, the manure is put into the pit. The compost preparations are then added – two sets per pit with 502 – 506 being put into five holes. The valerian preparation is diluted and activated by stirring it for 15 mins, using around 75 ml (2½ fl oz) in 2 l (67 fl oz) of water. The valerian preparation is sprinkled by hand on the completed pit. The pits are then covered with a wooden lid and left alone for eight weeks. The material is then turned in the pit every four weeks to keep it aerated. Once it has fully matured, it is used in compost teas, with seaweed spray and added to horn manure preparation (500) during the last 20 mins of stirring. The CPP takes four months to ripen in summertime and six months in wintertime. It is available to buy for NZ$7.50 (£4.00, $5.00) per 100 g (3½ oz).

CPP is the main way the compost preparations are applied to the land at the Seresin Estate. Analysis has revealed the presence of the so called *Bacillus subtilis* in the CPP, an organism that is antagonistic to botrytis and which is also included in commercial fungicides. There has also been less mildew on the Seresin Estate since spraying CPP – but Colin is not sure if CPP is the only reason.

The shed and pits for producing CPP.

Compost teas

The use of compost tea at the Seresin Estate is an important part of the vineyard management practices. The recipe varies – a common mixture consists of 1,200 l (312 ga) of water, 70 kg (154 lb) quality compost (at least one year old), 5 kg (11 lb) of CPP, 5–10 l (1–2½ ga) of worm-compost liquid, molasses, dolomite or basalt minerals and seaweed tea. Since compost, CPP and seaweed tea have already been inoculated with the compost preparations, these play an important role in the compost teas. The mixture brews while it is being stirred and aerated by a machine for 24 hours. It is then applied, undiluted, using the horse-drawn spraying machine to cover both plants and soil. The aim of the tea is to bring a multitude of living organisms into the plantations and, in so doing, suppress pathogenic organisms. It is not given as a fertiliser but as a tonic. For Colin, the purpose of compost tea is for 'founding life' since 'life begets life'. By giving the plant a dressing of living microbiological organisms with the tea, an explosion of pathogens is suppressed.

Seaweed tea

Kelp from the nearby coast is used for producing seaweed tea. It is put into a brewing tank without any water. It sits there for about three days and the juice dripping out at the bottom is collected. Afterwards, the tank is filled with water and the compost preparations are inserted by hanging them down, wrapped in a cloth, in a basket that is made to sink. One week later, the mixture is stirred and left to break down for several weeks. It is then used as a fertiliser. It too is sprayed using the horse drawn spraying machine. The seaweed application is not seen as a microbiological additive, but as an enzymatic and nutritional input.

Worm compost liquid

There are several tanks filled mainly with horse dung that function as worm farms. The compost preparations are added to the tanks. At the bottom of the tank, worm compost liquid is collected. This liquid is also used as a plant tonic and as part of the compost tea. Colin believes that worms and micro-organisms are sensitive to the radiation of the compost preparations and that these make them more active.

Summary

On the Seresin Estate, animal husbandry, grassland management and vegetable production is integrated into the vineyards and olive groves – the farm's main line of production. It is not only natural diversity that is celebrated there, but also social diversity. People from many different countries work together, share in and contribute to the 'Seresin Spirit'. Working with rigour, joy, energy and unrelenting curiosity is the vision of the owner Michael Seresin, and Colin Ross, as the estate manager, has brought this into the everyday life of the estate and made it the work ethos. This joyful and social approach to work on the estate is part of the quality Michael Seresin and the staff want to see flowing into the produce of the estate. Part of this approach has been to hire staff that are passionate about what they do. Special posts have been created, such as the post of 'preparations manager' held by Wendy Tillman.

Colin Ross, estate manager from 2006 till 2015, converted the farm to biodynamics. He experiences the earth as a living being and believes it to have its own intelligence as well as a memory. The management practices, including the pruning system and the spray applications, are intended as a means whereby the vines and the land can accumulate memories and so learn to self-organise and become a farm individuality.

Colin has the deep-seated goal of furthering life, and his actions and decisions on the Seresin Estate are everywhere shaped by this motivation. Having been a surfer and practising martial arts before coming to the wine industry, Colin has the experience that one has to be very positive and trust that one's intentions will succeed. This understanding flows into a positive approach of furthering life rather than fighting disease, and of applying the biodynamic preparations and other applications with full trust. The understanding that all beings of nature are living, spiritual entities is shared by Wendy Tillman, who is in charge of preparation work. Wendy has an intuitive connection to the vine and to biodynamics. Although coming from a scientific background, she doesn't need a scientific proof for the effectiveness of the preparations, since she follows her feelings and trusts her intuition in working with them. Both Wendy and Colin have observed the benefits the preparations have brought to the plants, animals and the land itself, as well as to the quality of the farm's produce. This evidence is enough to convince them of the efficacy of their preparation work and no rational explanations are considered necessary. The positive effects perceived are not only attributed to the preparations alone, but seen as a result of biodynamic practices combined with the intentions of the growers.

At Seresin Estate preparation work is part of social life. Making the preparations, and most of the stirring and applying, is a social event and performed by the staff as well as by biodynamic farmers of the region and other interested people. Due to her role as preparation manager, Wendy organises the preparation making and takes care of quality aspects and administrative duties, like record keeping for the Demeter certification.

Colin and Wendy both feel inspired by practical workshops and Peter Proctor's book *Grasp the Nettle,* but have also adapted their work with the preparations in response to practical challenges and the individual needs of the Seresin Estate. One special practice is to leave the stuffed organs to 'cure' for some days before burying them. Preparations are stored below ground in old wine barrels placed in the wine cellar, well protected from the danger of rats and floods. Derivative preparations and several teas complement the preparations. There is, above all, a large-scale production and use of the Cow Pat Pit preparation – the main way to get the compost preparations out onto the land. A special horse-drawn spraying machine is used to apply some of the preparations and their derivatives.

15. Binita Shah, Supa Biotech (P) Ltd, India

David Steiger, Dr Reto Ingold and Anke van Leewen

Introduction

Binita Shah is a landowner and farmer in Supi village, in the Nainital district of Uttarakhand, a steep and mountainous region at the foot of the Himalayas. She is the leader of a project that promotes biodynamic agriculture and is one of the main producers of biodynamic preparations in India.

Trained in biodynamic agriculture by Peter Proctor – a biodynamic farmer, carer and consultant from New Zealand – Binita has played a central role in spreading biodynamic agriculture in India. Her consultancy enterprise, Supa Biotech (P) Ltd, cooperates with the agricultural authorities and offers direct support to local farmers. There are 110 farmers in the valley below Binita's farm and, in the wider area, 207 small farms are currently working with biodynamic agriculture as a result of the efforts of Supa Biotech (P) Ltd.

The team from Supa Biotech (P) Ltd offers training in sustainable agriculture and extension services for the development of biodynamic agriculture in several states of northern India like Uttarakhand, Uttar Pradesh, Madhya Pradesh, Karnataka, Maharashtra, Punjab, Mizoram, Nagaland and Sikkim. Since it is not possible for each small farm to produce their own preparations, Binita set herself the goal of supplying preparations to the farmers of her region. The preparations are produced on a large scale and then sold or offered to farmers, development programmes and other initiatives.

For the last ten years Supa Biotech (P) Ltd has been officially recognised by the Biodynamic Association of India (BDAI) as a producer of biodynamic preparations. It works to the standards developed by David Hogg, a well known consultant of the Indian Biodynamic Association (BDAI). Binita's preparation work is inspected and assessed by IMO India, an independent certification body for organic agriculture. Kurinji Organic Farms in the

South of India is another producer of preparations and that is directly inspected by BDAI itself.

From November 6–9, 2015 Reto Ingold visited Binita Shah on her farm. Anke van Leewen, who was meant to accompany him, fell ill just before the trip and was not able to join in the visit to Binita Shah. On the first day of the visit, Reto Ingold was welcomed by Binita's farm manager, Mr Ramesh, and his co-workers. They showed him the different areas of the farm and the infrastructure of the preparation-making enterprise. On the second day, Binita Shah was also present and an in-depth interview about her work with, and relationship to, the preparations took place.

Set of compost preparations as produced and marketed by Supa Biotech (P) Ltd.

Farm portrait

Supa Biotech (P) Ltd (henceforth called 'Supa' or 'the project') is based in Nainital, in the state of Uttarakhand in the north of India, not far from the border with Nepal. The project is located in a rural area, where the average farm size rarely exceeds 2 ha (5 ac). The 5 ha (12 ac) farm of Binita Shah lies about 2,400 m (7,875 ft) above sea level in the first foothills of the Himalayas but still close to the plains.

The temperate climate of this region contrasts with the subtropical and tropical conditions in other parts of India. 'We have snow, we have spring and we also have the monsoon stuff in between,' said Binita. There are records indicating that in pre-colonial times the climate in the region was cooler. At that time the land was mostly used as common pasture land.

Then British colonists came, took the land and planted apple trees. Binita's grandfather bought the orchard from a British owner. There was no road then and it took him a whole day to walk from Nainital, to his orchard. Binita's father was in the army and never stayed on the farm. When he retired he moved to Nainital but by then he was too old and didn't have the resources to plant new trees or pay the workers. As a woman in India, Binita would not normally be entitled to inherit the property and her father had to write her into his will. Apart from fruit trees, Binita has also got some goats and a cow, and grows potatoes, kidney beans and the preparation plants.

Binita's farm is set up primarily to produce preparations and for educational purposes. She built a small hall called Proctor Hall for storing and packing the preparations. The Pfeiffer Hall is the room used for running courses and seminars.

Binita Shah checks the quality of a batch of the valerian preparation (507).

Supa works intensively with local farmers. They try to develop a market for Demeter certified products and take steps towards certifying local farmers. The farming system of the region is very diversified. Farmers work with horticulture, animals, bees and forestry, all in a very small space.

Today, Binita employs about twenty people on her farm. Their salaries are financed by the preparation business and the courses offered by Supa. The team is able to operate independently of Binita's direction. All the work is highly structured, everybody knows what to do and how to do it even when 'the land lady' is not around.

The forest

The forest plays an important role in the farming system. The women collect dry leaves in the forest and pile them up close to the livestock. The leaves are used for bedding and, once it has been composted in heaps, it is used to fertilise the potatoes and fruit trees. Most of the farmers keep cows or goats and they also take fodder for their animals from the forest.

View from Binita Shah's farm over the valley of Supi Village.

In March, the farmers are busy planting of potatoes as it is the main crop of the region. Then the fruit trees starts to flower. First the plums, followed by apricots and peaches and then the apples. The chamomile, valerian and other cultivated flowers only come into bloom in July. The second part of summer is marked by heavy rainfall, so the potatoes have to be harvested before July 15. After the monsoon season, in late autumn, the weather is dry. Most of the trees, except the evergreen species, lose their leaves. Around Christmas it usually snows.

From bottom to top, the mountainous hillsides have been made into terraces to protect their steep slopes from erosion and the leaching of soil. Nearly 27 tonnes (30 US tons) of top soil are lost per hectare (2½ ac) each year due to the torrential rain. Every winter, the terraces damaged by the monsoon have to be rebuilt.

First steps in preparation practice

As a child, Binita used to spend her holidays with her grandfather on his orchard in Nainital. She felt a deep connection to the land and already, as a 12-year-old girl, knew that one day she wanted to come back and settle in this place.

As a young woman Binita lived in Indore, Central India. Her interest in agriculture led her to a commercial forestry company with a livestock department, where she was able to gain practical work experience. Binita did not have any agricultural degree. 'In fact, I did not go into higher education because I did not want to be institutionalised,' Binita remembers. She wanted a practical experience of agriculture and this company provided a suitable entry point into this field.

In 1994, Binita had her first contact with biodynamic agriculture when she attended a conference led by Peter Proctor. She was bowled over by his lecture and her interest grew into an irrepressible desire to learn more about biodynamic agriculture. She approached him after his lecture and for the next three days they continued having lively conversations. On the fourth day, Binita realised that there must be something else behind biodynamic agriculture: 'He looked at me and said, "You are very switched on." and added, "Yes, there is a huge philosophy behind it. It is called anthroposophy." I think that was a watershed for me. I realised that this was exactly what I was searching for and I could spend my whole life on it. It fulfilled all my needs, the spiritual part, the physical part and the practical part, and I knew I wanted to professionally engage myself in a field like this.' From that day on, Binita started to study biodynamic agriculture, learning from the literature and also from Peter Proctor whenever he visited Indore. Binita recounted: 'We were sitting at railway stations and airports and all the while we were talking, he was teaching me. I have his handmade notes. While sitting at a railway station, waiting for a train, he would quickly draw the nitrogen cycle and say, "Look, this is how it works."' Binita invited Peter Proctor to come to Indore on behalf of the company for the next three years and asked him to teach about biodynamic agriculture.

Binita was able to convince the manager of the commercial forestry company where she was working at the time, to take up this subject and produce biodynamic preparations. Four years later, the company had invested a lot of money in it and urged Binita to set up a commercial operation. But Binita did not agree with the company's marketing model and realised that: 'I must step out and do it in my own small way.' The company continued without her and tried to commercialise the preparations, but already one year later 'the whole thing fell off.'

In 1998, when Binita Shah was 29 years old, she decided to return to her grandfather's farm. He had been well-known as a respected and wealthy businessman. Because of this, Binita received a lot of support from the local community when she returned. It was a region in which young people were rarely willing to take on their family farms. When Binita took on the farm most of the fruit trees were old. While she set about renewing the orchard, she also developed her business. Her focus was on knowledge sharing: her idea was that farmers would buy the preparations and in return they would receive free training on sustainable agriculture. From 1998 to 2002, Binita renewed the infrastructure of the farm and worked hard to build her own house and the facilities for making and storing the preparations.

For the next twelve years, Binita had to manage the farm and the continuing production of preparations from a distance, due to her taking up consultancy work for the government of the state of Uttarakhand. She explained that working for the government was necessary in order to give recognition to sustainable agriculture in her own and other states of India. She explained that: 'To work on a government platform that supported organic and biodynamic agriculture was a once-in-a-lifetime opportunity at a time when organic agriculture was still being snubbed, especially by the state agriculture university.' During this period of political involvement 'a number of policy decisions were taken that created an accepting environment for organic and biodynamic farming in all walks of life – among farmers, in industry, with consumers and in the scientific field. If I had missed this opportunity, the combined effort of the chemical industry and the university may well have succeeded in making the organic agriculture project a mere theoretical experiment. In many ways, the Uttarakhand model has influenced policy across the entire country, Sikkim followed Uttarakhand and succeeded in converting the entire state to organic practices, and Bhutan followed. More than ten states have organic farming policies and the entire country has become aware of the issue. It is not easy for the chemical companies to go about business as usual.'

In June 2015, after twelve years working as a consultant for the agricultural ministries of different states and in different projects, Binita returned to her farm with the aim of once again dedicating herself full-time to her farm and promoting biodynamic agriculture. She explained: 'For me, biodynamics is a part of life, it's an extension of my life. So I have to be completely part of it. I have to fit it somewhere in my culture and my way of understanding life… luckily it is all fitting in very well.'

How the work developed

The most influential person for the development of Binita's preparation work was Peter Proctor. When Peter first visited India in the early 1990s, he had no experience of how the preparations work in the tropics. Establishing the production of biodynamic preparations in India was also a learning process for him. For Binita, it was amazing to observe how Peter Proctor worked in new surroundings.

Peter was surprised by the efficiency with which plant materials are decomposed in a tropical compost. Binita noticed a difference between his composts and other composts she had seen. The quality of Peter Proctor's biodynamic compost was so much better. This was sufficient to convince Binita about the effectiveness of the biodynamic preparations.

Peter visited Binita's farm in 1996 for the first time. He identified the wild yarrow and dandelions growing in local meadows. They collected all the dandelions they could find and planted them together in a field. They took material from the local oak tree, went to the university library of Nainital and identified it as *Quercus dilatata*. Peter took a sample back to New Zealand and made a chromatogram. This convinced him that this species of oak was suitable for making the oak bark preparation.

When Binita moved to the farm in 1998, she immediately started making small amounts of the compost preparations. 'I brought some dried herbs, some chamomile and yarrow with me from Indore, 10,000 earth worms in a box, my dog, and a truck full of my luggage. At that time I didn't even have the house. I was living outside in a tent for two months and we kept the earth worms by the fire place, because I was afraid they would die.' During this time, she looked all around the region for a source of horns and finally came upon a 'trenching ground'. This is a place where dead animals are collected, useful parts taken out and the rest disposed of. It is work done by people from a 'low caste'. Binita obtains all the animal organs and horns from that trenching ground. She explained how it was culturally difficult for this practice to be accepted, since 'working with a dead animal and opening a dead animal is only done by a certain class of people in India. An upper caste will not even touch dead animals. So initially I did have some problems, but then slowly it was overcome.'

The first time she went to the trenching ground, Binita had a very significant experience. It was one that taught her how the animal organs can be connected with the cosmic forces. She observed a dead cow with its abdomen bloated with gases. She remembers: 'When I looked into that stomach, I felt as if I was looking at the world itself. I could see the whole universe inside this stomach ... This whole thing of using the preps and using the intestine and the mesentery, it became very clear. I could even relate the whole world inside the cow's stomach with the universe outside.'

All the required flowers and plants used for the compost preparations grow naturally in the region, some of them are native varieties but some of them only exist in very small numbers. Binita introduced new seeds of yarrow, chamomile, dandelion and valerian and cultivates the plants in fields.

Binita Shah explains her experiences with flow forms.

Binita's ongoing presence on the farm during those first years of making the preparations enabled her to adjust her approach to local conditions. When making horn manure preparation, for example, Binita learned how to cope with the abundance of earthworms, which one year destroyed the entire batch of horn manure. The worms had even gnawed into the horns and left marks. She decided to place the horns upside down and ensure that the gaps between the horns were filled with compost so as to distract the worms from the manure contained in the horns.

Binita's team developed many different techniques for improving the quality of the compost preparations. For example, they had to find the right way of stuffing chamomile flowers into the intestines 'so that it became neither too hard, nor too loose.'

The way oak bark preparation is made demonstrates how Binita has adapted her approach to suit conditions in the Himalayas and also meet specific practical challenges. Six years ago, Binita produced a large amount of oak bark preparation on her farm. She buried the 500 skulls in large ponds and set up a system of running water. But there was putrefying bad smell and in the summer she had to deal with an invasion of horse flies. They then made an interesting discovery – they sprayed CPP over the site and 'in just a week' the smell disappeared. The place where she acquires the cow skulls, however, is about 100 km (62 miles) from her farm. The transportation of fresh skulls became problematic and the highway authorities were getting suspicious. In order to avoid legal difficulties, Binita decided that instead of bringing the skulls to her farm, she would rent a space close to the trenching ground and produce the oak bark preparation there.

When she made the first experiments with flow forms in Indore, Binita had a profound experience. Peter had added some white cement powder to the water in order to make the flow of the water more visible. They spent a lot of time observing what was happening. When they switched off the pump, deposits of the white cement settled on the floor of the flow forms. They were surprised to see how the deposits traced the shape of the swastika symbol on the two sides of the forms. For Binita, the symbol of the swastika has a profound meaning. 'It is our sign for good things … whenever you enter a new house or anything, we put the swastika on the wall. It is the sign of continuity.' Another symbol closely related to the water movement in flow forms is the lemniscate, the symbol for infinity. When the water runs through the forms it clearly traces a sideways figure-of-eight, which is actually the logo of Supa Biotech (P) Ltd. Binita finds that this 'is the most amazing of nature's symbols … when you drive up along the Ganga river you can see that the river is creating that kind of form'. For Binita, using flow forms for stirring is a valuable method for use on large-scale farms. Most of the farmers in her region though are small-scale farmers and stirring is done manually.

Following a course given by Hugh Lovel, Binita has started making a horn clay preparation on her farm. In response to consultancy visits, she has also modified the way she prepares quartz for the horn silica preparation. Over the last four or five years, Binita has been making the silica powder even finer than she had originally learned from Peter Proctor.

The overall approach to making the preparations, however, hasn't changed from when Binita first learned it. She explained: 'Over the years we have made some learning progress. But I have not deviated from what Peter told me or what I learned from him, not much. It's not that I have started to do something on my own, I have mostly just followed it.'

Binita Shah's understanding of the preparations
Combining scientific evidence and spiritual knowledge

Binita tries to understand the preparations by drawing on the knowledge of conventional science and connecting it with her understanding of spiritual chemistry, for example. Finding evidence for the exceptional qualities of the preparations using conventional scientific testing methods, is what she uses to convince scientifically trained people of the importance of the preparations.

With regard to the compost made using the compost preparations, Binita has tested them for various substances and found significant differences. Biodynamic compost samples from Indore had a nitrogen content that was eight times higher than ordinary compost. Binita also tested biodynamic composts and CPP in terms of the carbon/nitrogen (C/N) ratio, nitrogen, phosphorous and potassium (NPK) levels, and micro-nutrients. It was especially with regard to a particular plant hormone that CPP gave impressive results. Binita says: 'There is no material in the world that has got a higher content of indole acetic acid, which is a plant auxin or hormone essential for plant growth, than the CPP.' Knowing the chemical substances present in a compost and in CPP assures Binita of their unique characteristics. Direct observation of their effects is, however, also important in convincing Binita that the preparations are essential for 'a natural form of agriculture'. Farmers connected to the project soon observed changes in their crops. 'Farmers were saying that there is a difference, that the potatoes were not drying out and also that the field is not drying out' Binita recounted.

Hugh Lovel, a chemist by training and inspired by Steiner, taught Binita about the importance of clay in biodynamic agriculture. As a result of conversations with him during his visits, Binita developed her own theory about how clay mediates between the 500 and the 501. Her understanding is that clay is the essential soil medium for ion exchange. The pH (power of hydrogen) decides the cation exchange capacity (CEC) of a given soil – a cation being a positively charged ion. Clay is an important constituent in the soil since hydrogen ions are able to act only on clay particles. Without clay in the soil there can be no ion exchange and the plants cannot assimilate nutrients. Since the exchange of ions is the most important process in plant nutrition, hydrogen has a very significant role – it mediates between the 500 (matter) and the 501 (spirit). According to Binita: 'Hydrogen is the mediator between the spiritual and the material, it is through the hydrogen that the spiritual world enters into the material world.' Hugh detected a clay deficiency in the soil of Binita's farm and recommended that she make an additional clay preparation to activate the clay processes in it. Hugh Lovel helped Binita to locate a place on her farm with a good deposit of fine clay. Tests of this clay showed that it is 90% pure clay, ideal for the 'horn clay preparation'.

Binita has also studied Hugh Lovel's articles on quantum agriculture and radionics, which deal with the question of marshalling the cosmic forces for plant and soil vitality. Binita is particularly inspired by Hugh Lovels' writings on the elements of the periodic table and feels these are very relevant for understanding the workings of the chemical elements.

Her main focus, however, is not on researching into the action-mechanisms of the preparations, but rather spreading practical know-how on how to use them, thereby making 'a natural way of farming' (integrating sustainability and spirituality) possible for the small farmers of India.

Using preparations to make a natural form of agriculture possible

Binita Shah is concerned about the state of Indian agriculture, the prevailing use of chemical pesticides, the dominance of multinationals like Monsanto, and national politics. According to Binita, the food most Indians consume in urban areas 'is not worth eating, it's just poison.' She added: '3,000 farmers are leaving farming every day in this country.' But Binita believes that biodynamic agriculture could be a solution for many young people and provide a motivation for them to return to rural areas, in the same way as she did.

Binita believes that biodynamic preparations are needed for the agriculture of India. The preparations provide a way to include and maintain a spiritual element in agriculture. In India, the average size of most farms is less than 2 ha (5 ac) and it would be too much to ask every family to make its own compost preparations. 'I know that this can change many things,' said Binita, 'but somebody has to make the preparations. The farmers need them. In India, not everybody can be expected to make the preparations, so I took on this role myself.'

For Binita, anthroposophical knowledge has no direct relevance to the ability of Indian farmers to work with the preparations. She argues that 'the Indian people are already burdened with spiritual teachings' and what they need are things that work in practice. Her main focus is therefore to provide the preparations and the know-how for using them so that they become part of a natural approach to farming. She also advocates making knowledge about biodynamics available to everybody who wishes to study it beyond the practical applications.

Biodynamic agriculture and Indian culture

Some aspects of biodynamic agriculture fit very well with Indian culture. Binita mentioned the sacredness of the cow and the understanding of planting calendars as examples of a cultural acceptance of biodynamic principles. A further congruence is the spiritual importance of Navaratri, the nine days of intense celebration and spiritual activity that follow the spring and autumn equinoxes in Hindu culture. Binita and some of her co-workers spend those special days fasting and trying to make at least some of the preparations during that time – such as filling horns with silica in spring and with manure in autumn.

There are other agricultural movements in India with some similarities to biodynamic agriculture, such as homa farming and Agnihotra practices. Binita and Binita's mother both practice Agnihotra, an ancient Vedic ritual consisting of the ritual burning of cow dung, rice and ghee in order to heal the earth.

The social setting

Technology transfer to small-scale farmers

While she was establishing herself on the farm, Binita tried to encourage local farmers to convert to biodynamics. She made use of an old Russian slide projector to show the farmers pictures of compost making at small meetings. This was the beginning of a social movement connected to her work with the preparations. 'Why would anybody listen to a twenty-nine-year-old woman, wearing jeans, having short hair and talking about agriculture, it doesn't make sense,' Binita said, remembering that the older farmers seemed to be more sensitive to her requests. The older farmers had grown up in a pesticide-free farming system and since her grandfather was respected as a 'big farmer' from a higher caste, respect for his leadership passed down to Binita once she had proved her loyalty to the local community. There is unspoken mutual support and respect stemming from current cooperation and past roles.

Interest in the biodynamic approach was growing and the state's department for agriculture started to pay attention to Binita's work with the farmers. There was a World Bank scheme to fund development projects in the region at that time. Binita managed to access the World Bank fund and run a development programme that focused on composting and enabled her to spread the word about biodynamic agriculture. She remembers: 'It was through compost that we brought biodynamics to people's attention.' Since compost and CPP were being used on different farms, many effects could be observed by the farmers.

Based on her experience with Peter Proctor, Binita also tried to show the farmers in the valley how they could improve their soil by using CPP and compost. Nor did she hesitate to instill them with fear by warning the farmers that their land and family would be destroyed if they continued using conventional farming methods. Little by little, and building on the goodwill the farmers still held towards Binita's grandfather, they started to listen and began implementing biodynamic practices.

CPP plays an important role in the spread of biodynamic agriculture. CPP can be produced easily and farmers can readily see the benefits of using it. Supa gives the compost preparations as a starter kit and teaches the farmers how to make CPP, which in turn is used to inoculate compost. This reduces costs and avoids making very poor farmers dependent on Supa. The guidelines do not include how to make compost preparations as that would be too complicated for them. 'We cannot teach all these traditional, small farmers to make the preparations, there is no point, it will not work here. But we can teach them to make the CPP, to make the liquid manures, and make the 500 and 501,' Binita explained.

Producing the biodynamic preparations now involves some close collaboration between Supa and the associated farmers. Forty-five farmers provide cow dung and in return pay a reduced amount to acquire the preparations. Their children also help to pick dandelion flowers and nettles on their way back from school. In return, Supa tries to achieve a good price for the farmer's products, supplies ready-made preparations to them and offers education and technical support. Many of the young men trained by Binita during the first years of her work found employment with the government department for agriculture. They work as specialists in organic and biodynamic farming, receive good salaries and spread the idea of sustainable agriculture.

Preparation practice

Supa produces all eight of the classical preparations given in the Agriculture Course as well as some additional supplements for biodynamic farming.

Field spray preparations
Obtaining and handling horns

In the Himalayan region, horns can be used for making preparations for at least for four or five years. This is different from other regions of India, where in the tropical climate the horns begin to disintegrate after their third use. Binita obtains all the horns in one year so that she doesn't need to replace them for some five years.

Horn manure preparation (500)

Binita's experience is that the quality of the cow dung plays an important role in the production of horn manure preparation. She believes that the quality of the manure depends more on the breed than on the fodder, even though fodder has an influence too. In India, both European and Indian cow breeds are kept. Binita prefers to use the manure of indigenous Zebu dairy breeds, such as Red Sindhi, to make horn manure preparation. Binita explains: 'The indigenous animal dung is very formed, you can actually break it in pieces ... even the fresh dung does not have any bad smell.' Binita's team has selected 45 farmers to provide the manure. Their cows have been inspected and examined for the quality of their manure. Those farmers have to make sure that the manure they provide for the preparations comes from lactating cows that are grazing and/or eating mainly fresh green or dried fodder but no grains.

The filled horns are placed next to each other with their openings facing downwards into beds. These beds form part of a terrace. The beds lie next to each other and are about 5 m × 5 m (16 ft × 16 ft). The spaces between the horns are filled with a mixture of soil and compost. A layer of soil is then put on top and the horns remain inside the earth for six months. The winter months tend to be dry, especially if it doesn't snow, and so the beds may need to be watered. Ten thousand horns are filled by hand and buried during the autumn season. Only the most well-structured manure is used and so can easily be filled into the horns by hand.

In spring, the horns are dug out and emptied next to the beds. The horn manure preparation is brought immediately to the storage shed. The store is made with rough brick walls surrounded by clay. It is a double-layered brick wall that is filled with topsoil from the farm forest. Binita couldn't agree to the import of peat and since she is too far away from any source of coconut fibre, she makes use of local forest litter instead. Moisture levels are maintained quite well in the store, but if the horn manure is stored for more than two years, it needs to be sprinkled with spring water 'just to keep it together'. The horns are cleaned and stored in the preparation storage shed over the summer.

Horn silica preparation (501)

Rough quartz crystals are collected from the Madhya Pradesh river valley. In spring, the crystals are broken into smaller pieces, initially with a hammer, then in a mortar and pestle, and finally between two glass panes to make the crystal particles as fine as possible. By adding water the powder is transformed into a paste. This paste is carefully filled into the horns. After a few minutes the top of the paste starts to dry out and seals the opening.

The silica preparation is buried in the soil from April to October, during the summer season of the northern hemisphere. The horns stay in the ground, open ends downwards to resist the wetness of the summer monsoon season. In September, the preparation is taken out and stored in glass jars exposed to the sunlight. About 30 kg (66 lb) of the 501 is stored this way. The idea 'is to make some every year just to keep the process going.' During 2015, only a few horns of horn silica preparation were produced.

Stirring and applying the spray preparations

On Binita's farm the preparations are mostly stirred by hand but there is also a 'Järna' and 'Slemstad' flow form that can be used to stir the horn manure and Cow Pat Pit (CPP) preparations.

Stirring by hand is done using a stick in plastic drums or buckets. The containers are filled with water, leaving about 20–25 cm (8–10 in) spare from the top so that a vortex can be created without the water overflowing. The vortex is fully developed when the base of the container becomes visible. At this stage they start stirring in the opposite direction. Three vortices should be created per minute.

The flow forms are used on Binita's farm about three times a year, whenever horn manure and CPP are applied to the whole farm. The water in the flow form is pumped round continuously in order to create a constant flow cycle. An electric pump with very low pressure is used. Binita is convinced that this is a viable alternative to stirring by hand on her farm. Binita can observe how the quality of the water changes as it moves through the flow forms. She says: 'The water becomes very colloidal, very oily.

It becomes thick, there is more weight, something expands.' But since she is not entirely sure about stirring with flow forms and has not been able to sufficiently examine the water's activity, she intends to continue stirring by hand as well.

The horn manure and horn silica preparations are stirred for exactly one hour. It is in fact quite a challenge to make the farmers comply with this, because stirring by hand is quite laborious. The challenge is to teach the activity of stirring as something that becomes part of the farmer's everyday life. 'Like a habit,' Binita explained. During training courses, stirring is often accompanied by a lot of singing and laughing, but the Supa team avoids adding any rituals that might distract from the activity of stirring itself.

The silica preparation is brought out early in the morning using knapsack sprayers that produce a fine mist. All the farmers in the valley keep a knapsack sprayer of their own for this purpose. For fruit trees there are foot operated pumps to spray the upper parts of the trees. These pumps have to be manually moved from tree to tree. The horn manure preparation is brought out late in the afternoon, and brooms made out of rice straw are used to sprinkle the preparation on the fields. This kind of broom is available in most parts of India and is normally used to brush down the whitewashed houses. By dipping the broom into a bucket of preparation and swinging it through the air a good distribution of water droplets across the field can be assured.

Binita does crop-related spraying of biodynamic preparations. Over the years she has developed spraying plans for different crops, such as basmati rice, sugar cane, onion, mango and apple trees. The sugar cane cycle, for example, takes nine months and the 501 is sprayed at least four times. Sugar cane has been found to respond well to the 501. For wheat, she recommends a first spraying of the 501 when the plant reaches the two leaves stage and a second when the grain's milky stage is reached.

Compost preparations

Yarrow preparation (502)

The only components of the biodynamic preparations that cannot be obtained locally are the stag bladders. There is a family of deer living in the farm forest, but in India hunting deer is forbidden. The stag bladders are collected, dried and pressed in New Zealand and sent to India each autumn in envelopes by Rachel Pomeroy, Peter Proctor's wife. Binita makes her decisions based on their ecological sustainability. She justifies the importation of bladders since they are the only ingredients needed for making the preparations that come from abroad and the dried bladders are easily transported.

Two types of yarrow grow on the farm: *Achillea millefolium* and a related wild *Achillea* species. The original seeds were collected and sown and now they grow up everywhere quite spontaneously. The single flower heads are picked during June, dried and stored for seven to eight months.

In April, the bladders are moistened, stuffed with yarrow flowers and hung up for four to five months. When the filled bladders are hung up in the sun over summer, they are protected with baskets. In the past the bladders were often snaffled by monkeys, who ran off and threw them away – 'just for fun,' said Binita.

In autumn, the bladders are put into clay pots and buried, and they stay in the soil during the winter season. In the beginning, Binita was concerned when they were excavated, that the floral structures of the yarrow were still visible and the colour wasn't as black as the flowers would become in a more tropical Indian climate. But she figured out that she simply needed to keep them stored in clay pots after they have been dug out. She found that: 'after a few months it looses that floral structure and it becomes black.'

Chamomile preparation (503)

Chamomilla matricaria is grown on Binita's farm. There are 'huge fields of chamomile. We plant them on air/light days and dry them. Chamomile is domesticated, it does not grow wild any longer, but it is Indian seed.' Flowering occurs as late as July. In autumn, the dried flowers are moistened and stuffed into the intestines, which are brought from the trenching ground. Then the filled sausages are placed in clay pots and buried. The preparation is dug out and stored in springtime. After a year in the store, it attains a very fine colloidal structure. This is the reason why Binita even prefers to use the preparations in their second year.

Nettle preparation (504)

Since *Urtica dioica* does not grow well on Binita's farm, she has resorted to using *Urtica parviflora* for producing the nettle preparation. According to Binita, *Urtica parviflora* works well as an alternative variety for producing the nettle preparation. Binita has planted enough nettles to meet the demands of her preparation production. The nettle leaves are harvested together with the upper parts of the stems. They are partially dry within a day of harvesting and are then put into clay pots and buried for a year.

Oak bark preparation (505)

Pieces of bark are cut from the *Quercus dilatata* oak trees growing in the neighbourhood of her farm, and ground 'completely like a flour'. Only a few fresh skulls are prepared on Binita's farm, the rest are made in the vicinity of the trenching ground where the skulls are collected. The cranial cavities of around 150 fresh skulls are filled each year with oak bark and then sealed with clay. The skulls are placed inside several plastic drums inside the building rented for the purpose. A supply of running water fed by gravity flows from drum to drum.

After removing the oak bark from the skull the preparation continues to mature in the clay pots during storage. Binita still buries about twenty goat skulls and two cow skulls in a small pond directly on her farm.

Dandelion preparation (506)

Binita started using three different varieties of dandelion. Two of them grow in the wild. *Taraxacum officinale* was introduced to the farm about five years ago and is now well established. But there are still not enough dandelion flowers growing on her farm to supply the amount of dandelion required for the production of sufficient dandelion preparation. Therefore, the local farmers help to harvest the flowers and bring them up to her farm. Binita asks them to pick the flowers together with the leaves. In this way, she can control whether they come from the correct species – there are other flowers that look very similar but are not dandelions. The flowers are harvested in June and July at an early stage of flowering. Binita underlines the importance of picking the flowers at the crack of dawn on an air/light constellation day. If the flowers are picked 'only one hour too late', she finds that the dandelion flowers open up their blossoms and 'convert into seeds' after drying. Flowers that have turned into seed are not used for making the preparation.

In autumn, the dried flowers are moistened and packed into the mesentery. If there is a need, Binita ties a cotton thread round them to hold the packages together, but normally they stick together skin to skin. The packages are placed into clay pots and surrounded by soil and compost and remain in the earth until spring.

Valerian preparation (507)

The European *Valeriana officinalis* was also introduced to Binita's farm. There is a native valerian variety, *Valeriana wallicii*, which grows widely in swampy areas, typically on the forests' edge or on the road side. This indigenous valerian is very well known in India for its medicinal qualities.

In Ayurvedic medicine the roots are used to heal epilepsy. They have a very strong and unique odour, similar to the European valerian, but much more powerful. Binita initially used only the wild variety to produce the valerian preparation, but it became difficult to collect enough flowers during its short flowering period from June to July. The propagation of the wild valerian, which would be necessary to obtain enough flowers, turned out to be very difficult, so Binita decided to import seeds of *Valeriana officinalis*, which is easier to cultivate on her farm than the native species.

As there are no other producers of valerian preparation in India, Binita holds an absolute monopoly and she is often asked to send her valerian preparation to biodynamic farmers all over the country. For the purpose of making valerian preparation, she even installed a dedicated room in her basement to store the large valerian juice bottles. The valerian preparation is made in that room so that high hygiene standards can be met and maintain the quality of this 'exclusive product'.

The procedure for making valerian preparation starts by harvesting the blossoms of flowering valerian. Some of the calyx can be included, but not the stems or the leaves. The blossoms are briefly put in an electric mixer to mash them into small pieces and 'to allow the juice to come out'. All the blossoms are put in buckets and completely covered with spring water. After leaving them to soak for eight to ten days, they are sieved through cotton cloth filters. The liquid is immediately bottled and the bottles must be filled right up to the top. The growth of pathogens in the airspaces can be avoided in this way and its 'wonderful sweet taste' can be preserved according to Binita.

Burying and storing practice

For both burying and storing the yarrow, chamomile, nettle and dandelion preparations, a receptacle is used made from two identical clay pots that fit together. Binita explains that: 'It's like an extension of the preparations themselves, it is rounded and very dark inside.' The clay pots have the important function of keeping earthworms out. The packages of yarrow, chamomile and dandelion inside the clay pots are surrounded with decomposed leaf litter from the forest floor. Once buried in their pits, they are covered with compost and soil. The nettle leaves are directly filled into the pots without adding any compost.

The preparations in clay pots are buried in the soil in beds of around 5 m × 5 m (16 ft × 16 ft). These are part of the farm's field terrace system. Afterwards, the finished preparations are stored in unglazed clay pots in

Proctor Hall. These pots are put in a container surrounded by peat. The finished preparations are always kept moist – or rather, they are prevented from drying out completely. Water can be added to the peat if necessary and this will penetrate through the pot into the preparation. Water is never added directly to the finished preparations.

The preparation storage in Proctor Hall.

Two identical clay pots are used to bury the stag bladders.
The spaces in between are filled with compost.

Derived preparations and other applications
Cow Pat Pit preparation (CPP)

CPP is produced in a 60 cm × 60 cm (24 in × 24 in) brick-lined cavity in the ground. Sixty kilograms (132 lb) of fresh cow dung from lactating cows is mixed with 250 g (9 oz) of ground eggshells, 250 g (9 oz) of basalt rock or bone meal, and 100 g (3½ oz) of jaggery (an unrefined type of sugar). When mixing is complete, the compost preparations are added. This mass is covered with a piece of fresh gunny cloth (hessian), and turned over three to four times before maturation is reached

The original idea of Cow Pat Pit (CPP) was developed by Maria Thun in the 1970s. Peter Proctor taught its production as a way of multiplying compost preparations and as an alternative to them in areas where compost was not applied. CPP has been successfully introduced to India where it is mainly used instead of the compost preparations for reasons of cost. Binita used to sell one set of preparation for 50 rupees (£0.50, $0.75), which is a low price, but some farmers were still not able to afford it. So the farmers started to use the compost preparation to produce CPP, which could then be used for further composting instead of the compost preparations. With a small amount of compost preparations a farmer can produce 40 kg (88 lb) of CPP. This allows them to produce a lot of compost and thereby multiply the power of the original compost preparations. Binita explained: 'We use the CPP as a preparation for compost. We are not replacing compost. We call it an extended inoculum.' But Binita would not like to see the compost preparations being replaced entirely by CCP. She supports the application of compost preparations in the making of compost for Demeter certified projects in accordance with Demeter International (DI) standards. Binita thinks that if a farmer can afford certification and can sell Demeter products, then he can also buy preparations to use in the compost. CPP has meanwhile found lots of new fields of application. For example, it is used widely to strengthen plants, ward off pests and to treat seeds.

CPP has to be stirred for at least 10 mins. It also can be added to the horn manure preparation during the last 15 mins of stirring.

Horn clay

The horn clay derivative preparation is made, like the 500, during the winter season. The clay is mixed with a little bit of water and then filled into the horns. The horns remain in the soil until springtime. Horn clay is used together with the 500. About 30 g (1 oz) of horn clay per hectare (2½ ac) are added to the horn manure preparation at the beginning of stirring in order to activate the clay processes in the soil. Binita uses horn clay only on her own farm and so she produces it approximately every third or fourth year.

Summary

The preparation work of Binita Shah is oriented towards the supply of preparations to small-scale farmers of Uttarakhand and throughout India. Her work with the preparations is done in association with forty-five farmers who provide some of the ingredients. Her enterprise, Supa Biotech (P) Ltd, helps to spread awareness for biodynamic agriculture through various projects and with support of the state government. Binita's project reaches out to over two hundred small-scale farmers working with biodynamic agriculture in the immediate surroundings of her farm.

Binita found her way into biodynamic agriculture through Peter Proctor. Her understanding and making of the preparations is influenced strongly by what she learned from Peter and what they have developed together in adapting the work with preparations to Binita's farm. In addition, she became interested in the approach of Hugh Lovel and, based on his consultancy, introduced horn clay to her farm.

Cow Pat Pit (CPP) plays an important role for the small-scale farmer and it is used as a way of multiplying the effect of the compost preparations. In this way, farmers can reduce their economic dependency and purchase fewer compost preparations.

Despite the large scale of the operation, Binita and her team at Supa produce the preparations carefully by hand, unaided by machines. Animal organ material is obtained from a 'trenching ground' where dead animals are cut up. The compost preparations (apart from oak) are buried and stored in unglazed clay pots. Stirring is done by hand and by using the flow forms. The preparations are stored moist in a building dedicated to this purpose – Proctor Hall.

Binita brings together in her work scientific laboratory testing, Indian spiritual traditions and anthroposophical understanding. Her main concern, however, is that the preparations are used for their practical benefits, because in her experience they greatly improve natural forms of farming.

Afterword

Dr Reto Ingold, Dr Ambra Sedlmayr, Ueli Hurter

It is our hope that these fifteen case studies will inspire practical work with the preparations and lead to new insights and further questions for preparation makers.

The case studies show how each of them is a jewel in its own right, shining in all its personal, social and environmental facets. There is an inner coherence to the portrayal of each individual approach to the preparations and this reflects the profound engagement of each interview partner. It is this inner coherence that we would like to highlight as a result of this study.

This result supports the argument that preparation makers need to have the freedom to follow their own very individual paths with the preparations. It is important that questions relating to preparation practice increasingly become the subject of reflection and open exchange within the biodynamic movement. This will hopefully lead to an open research-focused attitude. Decisions on minimum requirements for certification should not dampen discussions and experimentation with the preparations; there must always be an open space for research and innovation.

The social and environmental context in which preparation work takes place needs to be considered if the practical details are to be understood. Working with qualitative case studies allowed for an open exploration of the work of the preparation makers, capturing the subtle tones of how their personal experiences and inner attitudes influence their preparation practice.

Although the case study methodology can present a clear picture of many interrelated features, the interview and case study visit itself can only give a snapshot of the work of the interview partners. Some specific aspects of the findings presented may, therefore, need to be put in context by taking a broader look or focusing on them from a different angle. This is one reason why we did not want to systematise the results and draw any fixed conclusions. They could not live up to the complex intentions and experiences of the interview partners.

The case study interview partners took a risk in presenting their personal insights and sharing their approach to the preparations on their farm. For an outsider, it is easy to make an intellectual judgement without taking account of the background history of the work. The vulnerability and exposure of the interview partners will hopefully be handled with due respect and care.

It is hoped that the current study will contribute not only towards understanding the great diversity in current practice, but also show how, by practically engaging with the preparations, significant and very important common ground can be created. It is hoped that this will prove stronger than any differences that may arise from the life experiences, priorities, perceptions, and circumstances of individual preparation makers. The preparations are like universal 'music' that can be interpreted according to the personality and outer circumstance of each preparation maker.

More essential reading for biodynamic growers

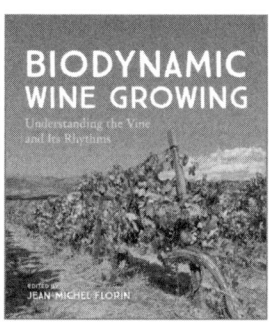

florisbooks.co.uk

For news on all our **latest books**, and to receive **exclusive discounts**, **join** our mailing list at:

florisbooks.co.uk/signup

Plus subscribers get a FREE book with every online order!

We will never pass your details to anyone else.